Mines of *Clark County*

by
William O. Vanderburg

This publication is a reissue of U.S. Bureau of Mines Information Circular 6964, originally issued in 1937, together with a reproduction of U.S.G.S. Bulletin 906-D, originally issued in 1939, authored by Eugene Callaghan

Published by Stanley Paher

Nevada Publications
Box 15444
Las Vegas, Nevada 89114

Acknowledgments

Thanks are due Frank Williams, Yeoman Briggs, and Albert C. Calkins, who gave assistance and information during the field work in the Goodsprings, Nelson, and Searchlight districts. respectively, and other owners, operators, and lessees too numerous to mention individually.

Charles White Merrill, supervising engineer, San Francisco office, Mineral Production and Economics Division, United States Bureau of Mines, furnished the data for the mineral-production tables, and the Mackay School of Mines, University of Nevada, at Reno, cooperated in making mineral determinations and assays and offering the use of its technical library.

Books in this series:

Mines of Humboldt and Pershing Counties

Mines of Clark County

Mines of Lander and Eureka Counties

Mines of Churchill and Mineral Counties

Mines of Goldfield, Bullfrog
and other Southern Nevada Districts

Mines of the Silver Peak Range, Kawich Range,
and other Southwestern Nevada Districts

Mines of Tuscarora, Cortez,
and other Northern Nevada Districts

Mines of Cherry Creek, Ely Range,
and other Eastern Nevada Districts

Mines of Battle Mountain, Reese River, Aurora
and other Western Nevada Districts

Mining Districts and Mineral Resources of Nevada

ISBN 0-913814-70-1

PRINTED IN THE U. S. A.

Foreword

In November 1937 William O. Vanderburg issued information circular 6964 on Clark County. It was one of a series of circulars dealing with mining and milling operations in various mining districts in the Western States. All were published under the direction of the United States Bureau of Mines.

The author set forth data on operating costs, grades of ore treated, wage scales, haulage rates, and information on mining properties from existing literature, mine operators, and local sources during the course of field work. This reprint of this circular gives us a picture of mining and milling operations in Clark County at the time of original publication.

The information in this book is reliable as to conditions at the time author Vanderburg compiled the original reports. Though later developments have brought many changes, the publisher and other interested parties recognize the importance of this report and decided to reissue it under the new title, Mines *of Clark County*. Three other books in this series give similar coverage of mines and mining in Lander, Eureka, Churchill, Mineral, Humboldt and Pershing counties (See back of book for details.)

TABLE 1. - Metal production, Clark County, 1908-1935.

Year	Gold		Silver		Copper		Lead		Zinc		Total value
	Fine ounce	Value	Fine ounce	Value	Pounds	Value	Pounds	Value	Pounds	Value	
1908....	13,553.03	$ 280,166	21,213	$ 11,243	62,719	$ 8,279	765,142	$ 32,136	1,115,851	$ 52,445	$ 384,269
1909....	16,594.41	343,037	35,294	18,353	23,308	3,030	445,907	19,174	3,013,352	162,721	546,315
1910....	10,603.36	219,191	30,650	16,551	216,773	27,530	1,309,684	57,626	2,707,071	146,182	467,080
1911....	1,229.40	25,414	55,643	29,491	185,915	23,239	1,629,571	73,331	3,548,032	202,238	353,713
1912....	2,401.43	1/ 49,740	236,465	145,426	173,887	28,692	6,555,565	295,000	13,254,860	914,585	1,433,443
1913....	6,970.35	144,090	198,922	120,149	351,855	54,537	6,265,783	275,694	14,369,709	804,704	1,399,174
1914....	4,468.74	92,377	127,233	70,360	188,154	25,025	4,207,419	164,090	11,862,149	604,970	956,822
1915....	8,109.54	167,639	160,749	81,500	301,327	52,732	4,655,985	218,832	21,061,182	2,611,587	3,132,290
1916....	14,912.66	308,272	404,788	266,351	573,088	140,980	8,405,796	580,000	29,400,537	3,939,672	5,235,275
1917....	10,599.88	219,119	319,836	263,545	878,646	239,870	9,373,817	806,148	20,660,549	2,107,376	3,636,058
1918....	8,391.95	173,477	252,694	252,694	467,486	115,469	6,541,129	464,420	15,472,155	1,407,966	2,414,026
1919....	7,016.45	145,043	258,288	289,283	148,469	27,615	4,257,114	225,627	5,974,219	436,118	1,123,686
1920....	5,870.69	121,358	186,945	203,770	121,508	22,357	4,018,864	321,509	9,381,503	759,902	1,428,896
1921....	3,602.78	74,476	12,094	12,094	31,247	4,031	284,217	12,790	69,397	3,470	106,861
1922....	1,668.21	34,485	11,682	11,682	12,899	1,741	360,500	19,827	891,174	50,797	118,532
1923....	1,093.03	22,595	36,669	30,069	24,005	3,529	1,519,764	106,384	8,755,427	595,369	757,946
1924....	1,838.59	38,007	127,182	85,212	23,257	3,047	7,764,771	621,182	7,909,042	514,088	1,261,536
1925....	3,867.48	79,948	78,418	54,422	22,610	3,211	5,409,440	470,621	6,230,664	473,530	1,081,732
1926....	3,275.86	67,718	39,121	24,412	21,240	2,974	2,044,599	163,568	2,210,162	165,762	424,434
1927....	1,286.39	26,592	15,843	8,983	9,879	1,294	1,690,443	106,498	3,546,078	226,949	370,316
1928....	749.52	15,494	27,805	16,226	118,085	17,004	2,220,320	128,779	1,425,285	86,942	264,445
1929....	1,566.29	32,378	21,022	11,205	47,000	8,272	1,080,457	68,069	1,236,702	81,622	201,546
1930....	1,368.77	28,295	6,565	2,527	18,346	2,385	966,686	48,334	2,474,323	118,768	200,309
1931....	5,659.39	116,990	12,662	3,672	8,887	809	796,107	29,456	1,899,328	72,174	223,101
1932....	3,100.93	2/ 66,419	8,592	2,423	4,353	274	450,752	13,523	254,795	7,644	90,283
1933....	1,866.83	3/ 39,424	5,038	1,763	4,201	269	75,631	2,798	23,346	981	45,235
1934....	3,906.26	4/136,926	19,783	12,789	10,352	828	154,437	5,714	28,635	1,231	157,488
	145,572.22	3,068,670	2,711,196	2,046,195	4,049,496	819,023	83,249,900	5,331,130	188,775,527	16,549,793	27,814,811

1/ Includes placer gold valued at $ 98.
2/ Includes placer gold valued at $2,317.
3/ Includes placer gold valued at $1,035.
4/ Includes placer gold valued at $402.

TABLE OF CONTENTS

Introduction to Clark County

Clark County, in southeastern Nevada, was organized and created from that portion of Lincoln County lying south of the third standard parallel south of the Mount Diablo base line, by an act of the State legislature approved February 5, 1909[1]. By this act the place formerly known as Clark's Las Vegas Townsite was renamed Las Vegas and made the county seat. The county has an area of 8,045 square miles, or approximately that of Massachusetts (8,266 square miles) or New Jersey (8,224 square miles).

According to the census of 1930, the population of the county was 8,532, the majority residing in Las Vegas[2]. The principal industries are agriculture, stock-raising, and mining.

The assessed value of real property for the fiscal year 1935-36 was $14,552,500. The county and State tax rate for that year was $2.52 per $100, exclusive of special taxes..

Topography

The topography of Clark County is characteristic of the Great Basin region and comprises a series of broad valleys between more or less parallel mountain ranges having a north-south trend. Most of the ranges are short; few extend in an unbroken chain more than 30 miles. The most prominent ranges are Charleston, Spring Mountain, Sheep, Las Vegas, McCullough, Opal Mountains, Muddy Mountains, and Virgin Mountains. In general the altitude is from 3,000 to 6,000 feet, although isolated peaks are higher. Charleston Peak, in the Charleston Range, the highest point in Clark County and the third highest mountain in the State, has an altitude of 11,910 feet; Sheep Peak (not to be confused with Sheep Mountain near Jean, Nev.), in the Sheep Mountains, is 9,706 feet high; Potosi Peak, in the Spring Mountain Range, is 8,500 feet; and Virgin Peak, in the Virgin Range, is 7,750 feet. Only a few of the mountain ranges have form sufficiently striking to give them individuality, and with the exception of the higher slopes in the Charleston, Spring Mountain, and Virgin Ranges are virtually barren of timber growth.

The altitude of the lower parts of the valleys ranges from 2,000 to 3,000 feet. The principal valleys are Indian Spring, Las Vegas, Muddy River, Piute, and the Virgin River.

Climate and Desert Vegetation

Clark County, with its limited rainfall and abundant sunshine, is one of the driest areas in the United States. During the summer the weather is quite warm, and a temperature above 110° in the middle of the day is not uncommon. However, the humidity is so low that a high temperature can be endured with less physical discomfort than with a lower temperature but higher humidity. In the valleys light snowfalls may occur, but the snow seldom stays on the ground more than a few days. In virtually all the districts mentioned in this report mining can be carried on the year around without difficulty.

Some kind of vegetation is prevalent over the whole area, with the exception of the lower parts of the valleys that have alkaline soil injurious to vegetation. On the higher mountain slopes there are growths of cedar, pinon pine, mountain mahogany, and some spruce. Among the common desert growths at lower elevations are the creosote bush (*Covillea tridentata*), greasewood (*Atreplex canescens*), sagebrush (*Artemesia tridentata*), Joshua tree (Yucca arborescens), and a great variety of plants from the Cactaceae family. The Yucca, sometimes mistaken for a cactus, belongs to the lily family.

The Cactaccae are particularly interesting, as they occur in greater number and variety in southern Clark County than in any other part of the State. Among the more common varieties are the fox-tail (*Coryphantha deserti*), California pin-cushion, (*Mammillaria tetrancistera*), the hedgehog (*Echinocereus mojavensis*), Indian strawberry (*Echinocereus Engelmannii*), beaver-tail prickly pear (*Opuntia basilaris*), flapjack prickly pear (*Opuntia chlorotica*), porcupine prickly pear (*Opuntia hystricina*), buckhorn cholla (*Opuntia asanthocarpa*), golden-spined jumping cholla (*Opuntia Bigelovii*), nigger-head (*Echinocactus Le Contei*), California barrel cactus (*Echinocactus acanthodes*), fishhook cactus (*Echinocactus polyancistrus*). and harem cactus (*Echinocactus polycephalus*). The barrel cacti are particularly interesting to the desert traveler, for many a life has been saved by the cooling, refreshing water stored within the pulpy interior.

Water Resources

The only large body of water in Clark County is Lake Mead, formed by Boulder Dam. When the water reaches the crest of the dam it will back up the Colorado River about 115 miles and have an area of 227 square miles[3]. The water supply for Boulder City is obtained from this lake; it is filtered and chlorinated for domestic use.

Water for domestic use in Las Vegas and for irrigation is obtained from artesian wells in Las Vegas Valley. The artesian basin has an area of 100 square miles, and in it there are 300 wells 300 to 1,100 feet deep. One of the wells owned by the Union Pacific R.R. is 650 feet deep and flows 3 1/2 million gallons per day.

The only areas in the county where enough water is available for irrigation are the Las Vegas, Virgin, and Muddy River Valleys. The Virgin River rises in the mountains in southwestern Utah and flows southwesterly across the corner of Arizona into Nevada, emptying into Lake Mead. Its principal tributary is the Muddy River, which has its source in Lincoln County, Nev. It flows almost due south, joining the Virgin River a little below St. Thomas, Nev. These rivers supply water to irrigate only small portions of the very fertile valleys bearing their names. The chief crops are cantaloupes, tomatoes, alfalfa, and various truck-garden products.

In the valleys, the precipitation is 2 to 3 inches per year. The amount of precipitation increases progressively with the elevation, and in the higher mountains it is considerably more. Cloudbursts that may do considerable damage sometimes occur during the summer months.

In the Goodsprings district, water for mining and milling is available from shallow wells; in the Searchlight district, from mine workings; and in the Eldorado district, from the Colorado River about 7 miles east.

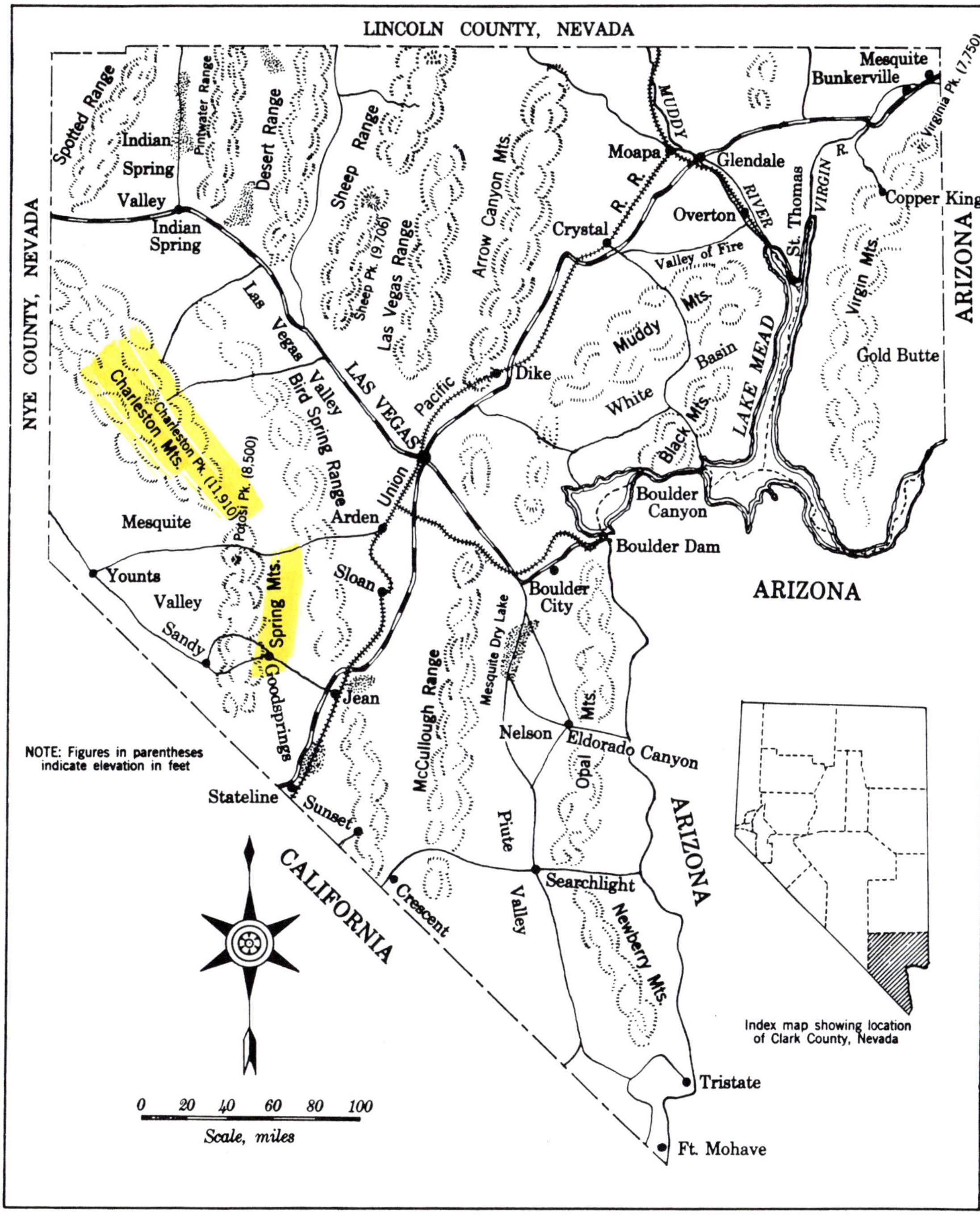

Figure 1.—Sketch map of Clark County, Nev.

Power Facilities

Clark County has no public-service power company at present, and all the mining districts depend upon either Diesel or gasoline engines for power. Work on Boulder Dam, on the Colorado River at Black Canyon, 25 miles southeast of Las Vegas, was begun officially on July 7, 1930, and was virtually completed in 1936. Installed hydraulic turbine capacity at the dam will be 1,850,000 horsepower, of which the firm output is estimated to be 663,000 horsepower. Eighteen percent of the firm power can be withdrawn for use in Nevada, subject to 6 months to 2 years notice on contract, depending upon the amount withdrawn.

The Southern Sierras Power Co. owns a 132-kv transmission line from the Dam to southern California, and the City of Los Angeles has two 278.5kv transmission lines to Los Angeles. A transmission line to Pioche, Nev., is now under construction.

A number of problems connected with the use of Nevada's allotment of power have not been worked out, and the interested reader is referred to the report of the Colorado River Commission of Nevada[4]

Transportation Facilities

Las Vegas is on U. S. Route 91 (Arrowhead Trail), which connects Salt Lake City, Utah, and Los Angeles, Calif. This highway is oil-surfaced and traverses the county northeast and southwest. Other oiled roads connect Las Vegas with Boulder City and points south, via the road over Boulder Dam; Las Vegas with Beatty, Nev., and points north; and Glendale, on U. S. Route 91, with Overton, Nev. All the other roads in the county are either gravel or natural and afford easy access by automobile to all the districts mentioned in this report, except the territory between the Virgin River and the Arizona-Nevada boundary. Here the roads are in poor condition and in places virtually impassable.

Western Air Express operates three planes daily between Salt Lake City and Los Angeles, stopping at Las Vegas enroute.

The trunk line of the Union Pacific R.R. connecting Salt Lake City with Los Angeles traverses the county in a northeast-southwest direction, roughly paralleling U. S. Route 91. A branch line extends from Las Vegas to Boulder City and another from Moapa to a point several miles south of Overton. The freight rates on ores from various points in Clark County to Salt Lake Valley smelters are shown in the following table.

Value of ore per ton	$15	$20	$30	$40	$50	$60	$70	$80	$90	$100
Roach	4.00	4.00	4.50	5.00	6.60	7.30	8.00	8.70	9.40	10.10
Jean	4.00	4.00	4.50	5.00	6.60	7.30	8.00	8.70	9.40	10.10
Sloan										2/9.75
Arden										2/9.75
Las Vegas										2/9.75
Boulder City					6.30	6.80	7.30	7.80	8.30	8.80
Moapa 1/					6.30	6.80	7.30	7.80	8.30	8.80
Arrowhead 1/		5.29	6.30	7.43	8.44	9.11	9.45	9.90	10.24	10.58
St. Thomas		5.29	6.30	7.43	8.44	9.11	9.45	9.90	10.24	10.58

NOTE: Minimum weight, 40,000 lb., unless otherwise specified.
1/ Minimum weight, 80,000 lb.
2/ Minimum weight, 60,000 lb.

The freight rates shown in the table do not apply to industrial minerals, since it is the practice of railroads to give special consideration to enterprises that can supply a large and steady tonnage, and such rates are determined largely by competitive conditions.

History of Mining

Nevada was part of the territory acquired from Mexico in 1848 by the treaty of Guadalupe-Hidalgo. In 1850 it was made a part of the Territory of Utah. By Act of Congress, approved March 2, 1861, the Territory of Nevada was created, and on October 31, 1864 it was admitted into the Union.

Clark County is one of the oldest sections of Nevada, and the first mining in this area was done by Indians who settled in the Muddy and Virgin River Valleys at the dawn of the Christian era. Archeological investigations at Pueblo Grande de Nevada (The Lost City) near St. Thomas show that these ancient inhabitants reached a fairly high stage of civilization. Their homes, burials, and specimens of their culture may be seen in the Lost City museum a short distance south of Overton. The mining activities of this aboriginal race were confined largely to the exploitation of the Virgin River salt deposits (probably for trade purposes) and for turquoise. Pipes and other articles made from soapstone have been found also.

The earliest white men in Nevada were the Franciscan monks, who passed through Clark County on their way from Mexico to California as early as the middle of the eighteenth century. There is some evidence to show that a little mining was done by Spanish explorers, who followed in the trails of the Franciscan fathers.

Southern Nevada was colonized by the Mormons in 1850, and settlements were established at Las Vegas, Callville, and along the Muddy and Virgin Rivers. Aside from the mines of the Indians and Spanish explorers, the Potosi mine in the Goodsprings district is the oldest lode mine in the State. It was discovered in 1855 by some of the Mormons who returned from San Bernardino, Calif. According to Helen J. Stewart[5],

> A man by the name of Slade was made superintendent of the Potosi mine in 1855, having been sent out by the church authorities to supervise the lead mining. They made an attempt to smelt the ore at the mine, using pitch-pine for fuel, with no result save badly burned hands. They also tried cedar wood for that purpose, which was better, but still not successful. Not being satisfied with the results, they brought their ore down to the Las Vegas rancho. Dudd Leavitt and Isaac Grundy here built a furnace in a fireplace, using the chimney for making a draft. When the ore became too hot they devised the plan of placing an adobe brick in the furnace to even the temperature. In this crude way they succeeded in making a success of their smelting operations. They moulded their lead in an old iron skillet which gave the bars the appearance of miners' loaves of bread. In this manner they prepared and sent to Cedar City, Utah, ten thousand pounds of lead, which was put in charge of Bishop Smith and by him distributed.

Notwithstanding this early discovery, little attention was paid to the mining of base metals until 45 years later.

The first systematic mining in Clark County began with the discovery of gold-silver deposits in the Eldorado (Colo.) district in 1857. As this district was on a natural travel route (the Colorado River). it was actively exploited. This discovery was followed by the discovery of the Keystone mine in the Goodsprings district in 1892 and mines in the Searchlight area in 1897.

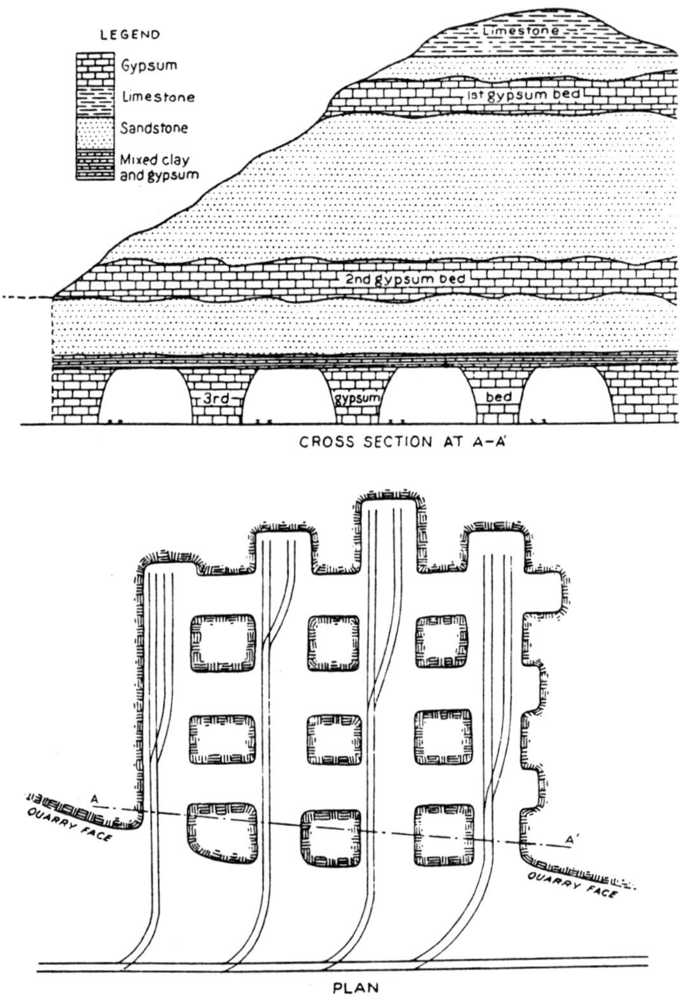

Figure 2.—Cross section and plan of room-and-pillar workings, Blue Diamond Corporation. Ltd., Arden, Nev.

In 1905 the San Pedro, Los Angeles & Salt Lake R.R. was completed and the town of Las Vegas established. The completing of this railroad stimulated the mining of zinc, lead, and copper deposits in the Goodsprings district, and the greatest activity occurred here during the war, when the price of base metals became very high.

The exploitation of nonmetallic deposits in the county began about 1910. In recent years most of the metal mining has been done by lessees and small companies, principally of gold and silver deposits.

Mineral Production

The annual metal production of Clark County from 1908 to 1934 is shown in table 1. Nearly all the metal shown in the table was produced in three districts; the gold and silver came principally from Searchlight and Eldorado, and the copper, lead, and zinc from Goodsprings. Other metals produced in commercial quantities are manganese, vanadium, molybdenum, cobalt, platinum, and palladium. Turquoise has also been produced for semiprecious gem stones.

The county has also produced considerable tonnages of certain industrial minerals. Production statistics of these minerals are not available, but from information gathered in the field past production is estimated as follows: Limestone and dolomite, 1,500,000 tons; gypsum, more than 1,000,000 tons; silica sand, 200,000 tons; borates (colemanite), 200,000 tons (B_2O_3 content about 20 percent); and small amounts of rock salt and feldspar. Large reserves of the foregoing industrial minerals occur in the county, and while some of the deposits have little commercial importance at the present time on account of transportation difficulties and distance from consuming centers, they constitute valuable sources of supply for future needs.

Extensive reserves of magnesite occur in the St. Thomas district, but the material contains too great a percentage of impurities for commercial use. To utilize this magnesite on a large scale as a source of magnesium or for other purposes, research work is necessary.

In February 1937 the writer estimates that 360 men were directly employed in the mining industries of the county; two-thirds of these were employed in metal mining and one-third in the production of nonmetallic minerals.

ALUNITE DISTRICT
(Gold)

The Alunite district is in central Clark County 3 to 5 miles west of Boulder City, the nearest railroad point. In 1908, gold was found here by Robert T. Hill, and the Alunite Mining Co. was organized to prospect in this area. The country rock is generally alunitized, and it was thought that because of the similarity in rock alteration to the Goldfield district important gold deposits would be found. Although considerable exploration was done, this company did not find anything of importance.

The alunite of potash variety is widely disseminated in the country rock as well as in small veins. The district has been examined on several occasions as a source of potash, but on the average the alunite is too low grade to be of commercial importance at present. Samples taken by Gale[6] indicate that the average K_2O content is probably 2-1/2 to 3-1/2 percent.

There is no record of production from this district, but the Quo Vadis property is reported to have made several small shipments of very high grade ore. This property was first worked in 1915 by a company called the Quo Vadis Gold Mining Co. In February 1937 a small company called the Boulder Canyon Gold Mining Co. was preparing to prospect the Quo Vadis property.

Quo Vadis Group

The Quo Vadis property comprises a group of five unpatented claims owned by E. W. Clark, of Las Vegas, Nev., and associates.

Development work consists of a 1 1/2-compartment vertical shaft 90 feet deep and an adit 300 feet long. Mining equipment includes a GardnerRix compressor (6 by b inches) driven by a Buick automobile engine, a 15-horsepower Fairbanks-Morse geared hoist, mining tools, and blacksmith shop.

Rich stringers of free-milling gold ore occur in biotite monzonite formation. The gangue is chiefly quartz and calcite stained with manganese. A small amount of silver is associated with the gold.

ARDEN DISTRICT (Gypsum, Silica Sand)

Gypsum deposits occur in the Spring Mountains 5 to 13 miles west of Arden, Nev., a station on the Union Pacific R.R. Gypsum was mined in this locality by the Arden Plaster Co. from about 1909 to 1919. In 1919 the holdings of the Arden Plaster Co. were purchased by the United States Gypsum Co., and the latter company operated until 1931, when the mine and mill closed because the deposit was exhausted. The plaster mill at Arden is still intact, but much of the equipment is either in poor condition or obsolete. In 1925 the Blue Diamond Corporation, Ltd., began mining a gypsum deposit 13 miles westerly from Arden. This company has operated steadily and in 1936 was the only active producer of gypsum in the district. Adjoining the Blue Diamond mine is the Mateucci gypsum deposit, which has never been exploited. This property is reported to contain large reserves of gypsum comparatively free from anhydrite and of excellent grade.

Statistics on the production of gypsum from the Arden deposits are not available, but judging from the extent of the workings production has probably exceeded 1,000,000 tons.

Unlimited reserves of silica sand suitable for molding purposes occur west of Arden. Molding sand has been produced on a small scale from this deposit by the Arden Silica Co. for several years.

Blue Diamond Corporation, Ltd.

The property of the Blue Diamond Corporation, Ltd., comprises about 1,000 acres, one-third of which is covered by a gypsum deposit. The company employs an average of 35 men in the production of about 300 tons of gypsum per day. The gypsum is crushed and screened at the mine and is shipped to southern California for use in cement and building industries.

The following data on operations were obtained from a report by Bradley[7] and from notes taken by the writer on a visit to the property late in 1936.

The main gypsum deposit occurs in a nearly horizontal sedimentary bed ranging from 12 to 20 feet in thickness and covered by strata of limestone and sandstone and irregular beds of gypsum. The overburden ranges from a few feet to 100 feet, and in the thickest portions the main gypsum bed contains masses of anhydrite. Directly overlying the gypsum bed is a layer of gypseous clay about 2 feet thick, which has enough strength to support the overburden and permit a room-and-pillar system of mining. The footwall is limestone and the parting between it and the gypsum bed is distinct. The gypsum analysis is as follows:

	Percent
$CaSO_4$	78.43
SiO_2	.65
$MgCO_3$	.25
$Al_2O_3;Fe_2O_3$	.02
H_2O	20.65
Total	100.00

Mining is done either by room-and-pillar or by open-cut methods, depending on the depth of overburden. In the open-cut system of mining the overburden is broken by blasting and shoveled into 5 1/2-ton-capacity dump trucks with a 3/4-cubic-yard-capacity Osgood gasoline shovel. The trucks haul the material to the nearest convenient dumping site, which is usually not more than a few hundred yards. The stripping depth depends upon the thickness of the gypsum bed; ratio of the gypsum mined to the overburden removed is, roughly, 1 to 1.

A sketch of the room-and-pillar method of mining is shown in figure 2. the rooms range from 25 to 40 feet in width and average about 12 feet in height. Pillars are left at intervals of about 35 feet. Drilling is done with jackhammers and 7/8-inch hollow hexagonal steel. About 28 holes, 8 feet deep, are required in each heading per round. Blasting is done with 40-percent gelatin dynamite and electric detonators. Air for drilling is furnished by two Chicago pneumatic compressors each 10 by 12 inches in size.

The broken gypsum is shoveled by hand into bottom-dump cars that hold 2.4 tons each. A small amount of waste, mostly anhydrite, is sorted out by hand. The anhydrite is readily distinguishable from the gypsum because of the difference in its color and texture. It is hauled to the crushing plant, about 1/4 mile distant, in trains of 8 cars drawn by a Plymouth gasoline locomotive. The mine is thoroughly ventilated, so that the exhaust gases from the locomotive are not an underground hazard.

The flow sheet of the crushing and screening plant is shown in figure The cars are dumped into a 30-ton storage bin that has a grizzly bottom with 3/4-inch opening for by-passing the fines to the conveyor belt. The oversize passes to a Dodge crusher (24 by 34 inches) set to crush to 4-inch size. From the Dodge crusher the minus 4-inch material passes to a rotary crusher (42 inches) set to crush to 1 1/4-inch size. The minus 1 1/4-inch product and grizzly undersize are conveyed by belt conveyor to a tramway bin of 100 tons capacity. From the tramway bin the gypsum passes to a measuring pocket regulated by hand, and thence into the aerial tramway cars each holding 1,800 pounds. The tramway is operated by gravity, the loaded cars taking the top cableway track and the empties returning inverted on the lower one. The tramway discharges to a set of stationary scalping screens. The screen products are diverted to loading bunkers and thence to railroad cars.

Power for mining and the crushing plant is furnished by a 180-horsepower Fairbanks-Morse, 3-cylinder, Y-type, Diesel engine.

The direct cost of mining is about $1 per ton, including crushing, screening, and loading, but exclusive of depreciation and depletion.

Arden Silica Co.

A deposit of silica sand worked by the Arden Silica Co. of Los Angeles, Calif., Harry E. Blood, president, occurs 4.7 miles westerly from Arden, a station on the main line of the Union Pacific R.P. This company has worked the deposit under a lease agreement for the past 4 years, and has produced an average of 7 carloads of molding sand per month. No attempt has been made to utilize the sand for glass making purposes. An average of four men is employed.

The silica occurs as well rounded grains in a cross-bedded sandstone formation at least several hundred feet thick. The sandstone is coherent but it is easily broken down to individual grains by crushing and washing. In places the silica is either stained with iron or cemented with lime.

The sandstone is mined by the room-and-pillar method. Rooms 12 feet wide and 10 feet high are driven into the hill for about 100 feet. Drilling is done with jackhammers and 7/8-inch hollow hexagonal steel. Blasting is done with 40 percent gelatin dynamite, No. 6 detonators, and tape fuse. Compressed air is furnished by a 120-cubic-foot-capacity compressor driven by a 4-cylinder gasoline engine. The sandstone is shoveled into trucks by hand and hauled several hundred feet to the crushing plant.

Crushing-plant equipment consists of a Blake-type jaw crusher (18 by 12 inches), a set of rolls (18 by 30 inches), a belt and bucket elevator, and an Isbell vibrating screen equipped with 14-and 35-mesh screens. The minus 35-mesh screened product is discarded as waste, and the plus 35-minus 14-mesh product is loaded into a truck and hauled 4.7 miles to the washing plant at Arden. Power for the crushing plant is furnished by a 75-horsepower gasoline engine.

Washing-plant equipment consists of a belt and bucket elevator, two screw classifiers, conveyor belt, and storage bin. The classifiers are spiral steel screws, 14 inches in diameter and 12 feet long, that operate in a box launder inclined 19°. Power for the washing plant is furnished by a 75-horsepower gasoline engine.

COPPER KING (KEY WEST) DISTRICT (Copper-Nickel-Platinum, Gold-Silver-Lead, Mica, Beryl, Manganese)

The Copper King district is in the west foothills of the Virgin Range about 15 miles a little west of south from Bunkerville, Nev. It can be reached by automobile by turning south off Highway 91 at Riverside and driving 1.6 miles to the Darling ranch, from which point a winding road goes to the Key West mine, 10.3 miles distant. The former Arrowhead Trail route between Bunkerville and St. Thomas traverses the district, but this road is impassable by automobile.

Copper ore was discovered on the Key West property in the late nineties by Scott Allen, a Cherokee. From 1900 to 1903 the property was explored primarily for copper and nickel by a company called the Nevada Copper and Nickel Co. During the course of this exploratory work, platinum was discovered in the ore by S. W. Darling, who was superintendent of the operations. Although considerable development work was done, very little was produced because of the metallurgical difficulties involved in recovering the metals from the matte.

After the Nevada Copper & Nickel Co. ceased operations, at least three other attempts were made to work the Key West mine, the last being made in 1936 by the International Smelting & Refining Co. of Salt Lake City, Utah.

From other properties in this area small operators and lessees produced (from 1908 to 1932) 1,550 tons of ore that yielded 51.39 ounces of gold, 1,313 ounces of silver, 99,990 pounds of copper, and 26,597 pounds of lead, valued in all at $19,726, an average value of $12.73 per ton.

Mica, boryl, and manganese also occur in this district but virtually none has been produced.

Key West Mine

The Key West group of seven patented claims is controlled by the R. E. Payne estate of Boston, Mass. The mine lies at an altitude of 3,665 feet. This property is unusually interesting because nickel and platinum are present in the ore, two metals that are comparatively rare in the United States.

The only record of production from the mine is one carload of ore (91,600 pounds) which was shipped to Fredericktown, Mo., about 1903, for testing. According to S. W. Darling, this ore was taken from various places in the mine in order to obtain a representative sample, and the analysis was as follows:

Silica	34.87 percent
Iron	10.59
Lime	9.41
Alumina	6.55
Magnesia	15.33
Copper	2.30
Nickel	1.79
Cobalt	.08
Sulphur	6.90
Platinum metals, per ton	.13 ounce
Gold, per ton	Trace
Silver, per ton	Trace

The sample contained a small trace of lead, no zinc, and no arsenic; it also contained 0.3 percent titanic oxide.

In 1936 the Key West and other properties in the vicinity were under option to the International Smelting & Refining Co. When the property was visited in November, this company employed a crew of 20 men to rehabilitate the mine. Equipment included a 120-horsepower Caterpillar Diesel engine belt-connected to a 500-cubic-foot-capacity, Sullivan, angle-compound compressor, an 80-horsepower Caterpillar Diesel direct-connected to a single-drum Sullivan hoist, and a 60-horsepower Caterpillar Diesel beltconnected to a Byron-Jackson, 350-gallon-per-minute-capacity, deep-well pump. After the mine was unwatered and a small amount of development work done, the option on the property was relinquished.

According to Bancroft[8], the formation in the vicinity of the Key West mine is gneiss, apparently of granite origin. The gneiss is intruded by basic dikes, which, in turn, have been cut by dikes of aplite and pegmatite. The ore bodies occur as lenticular masses in the most basic of the dikes, which are a variety of peridotite. These peridotite dikes are interesting as they constitute one of the few examples of pre-Cambrian ore deposits in Nevada. The strike of the Key West dike is north 65° east, and the width varies from 8 to 60 feet, averaging about 25 feet. It is cut by numerous faults. To the southwest of the mine the dike is covered by a mantle of detritus, but it can be traced to the northeast, where the topography is rugged for several miles.

In the sulphide zone the ore occurs either as a massive or disseminated sulphide consisting principally of pyrite, chalcopyrite, and pyrrhotite, the latter probably nickeliferous. The platinum appears to be uniformly disseminated through the ore, but the mineral has never been determined.

Great Eastern Group

The Great Eastern group of seven unpatented claims lying 1 1/2 miles northeast of the Key West mine is owned by S. W. Darling and A. M. Thompson of Riverside, Nev. This group was located about 35 years ago by the present owners, who did considerable development work by hand methods. In 1936 the property was under option to the International Smelting & Refining Co., but it reverted to the owners about the first of 1937. It has never shipped any ore.

Development work comprises three main adits ranging from 400 to 600 feet in length. Total workings comprise approximately 1,800 feet. There is no equipment on the property.

The peridotite dike has been cut by the three adits, and the geology is similar to that of the Key West mine.

Other Copper-Nickel-Platinum Claims

The United Nickel group of 13 unpatented claims owned by Robert A. Baker and associates, of Las Vegas, adjoins the Great Eastern group on the northeast. Two unpatented claims owned by Robert A. Baker individually adjoin the Key West group on the south.

In 1936 the International Smelting & Refining Co. is reported to have located a group of 32 claims near the Key West mine.

No real development work has been done on any of the foregoing claims.

Mica and Beryl Claims

On the western slope of the Virgin Range near the Arizona-Nevada boundary line about 12 miles east of Mesquite, Nev., muscovite mica and beryl occur in pegmatite dikes. The deposits are reached by a poor automobile road, which connects with Highway 91 at the west end of the bridge over the Virgin River near Mesquite. From the bridge the road winds a little east of south for a distance of 9.6 miles. The last 1 1/2 miles must be traversed on foot over a steep trail. Only a small amount of sheet mica has been produced, said to have been shipped some years ago. In recent years there has been no activity.

B. R. McKenna, of Moapa, Nev., and partner own two unpatented claims. On these claims a pegmatite dike has been explored by at least five shallow shafts and open cuts. The dike on which most of the work has been done varies from 1 to 3 feet in width, with a nearly vertical dip. The formation is mica schist. The minerals in the dike are mica, quartz, and feldspar, with occasional crystals of tourmaline and beryl, the latter up to 1 inch in diameter. Insufficient prospecting has been done to determine whether or not beryl occurs in quantities of commercial importance. The mica occurs in books up to several inches thick and from 6 to 8 inches square. The mica seen by the writer contains specks of foreign material and is not cleavable in large sheets.

The Fools Gold group of five unpatented claims recorded in Mohave County, Ariz., is owned by Lloyd Williams, Clarence Stay, and associates, of Las Vegas. According to Stay, crystals of beryl up to 3 inches in diameter have been found on this group of claims in a series of pegmatite dikes up to 12 feet or more in width.

Manganese Deposit

There is said to be a deposit of manganese in the Virgin Range 18 miles southeast of Mesquite, Nev. The last 4 miles to the deposit must be traversed either on foot or horseback. The deposit is covered by three unpatented claims recorded in Mohave County, Ariz., and owned by Ernest A. Walker and associates of Mesquite. According to Walker, picked samples of manganese ore analyzed 30 to 40 percent manganese, but on the average the grade is considerably less. The deposit has been prospected by a number of shallow open-cuts, but no manganese has ever been produced.

CRESCENT DISTRICT
(Gold, Silver, Turquoise, Feldspar, Fluorspar, Mica)

The Crescent district is in southern Clark County near the California-Nevada boundary line 6 miles east of Nipton, Calif., a station on the Union Pacific R.R. The Sunset district adjoins it on the north and the Vanderbilt district, in San Bernardino County, Calif., on the south. Prospecting began in this area about 1894, but no important discoveries were made. The period of greatest activity was from 1905 to 1907, when at least 10 incorporated companies were Working in this area. During this period the little settlement of Crescent, 6 miles east of Nipton, was established, but it was abandoned soon after. Late in 1936 metal-mining was revived, stimulated by the increased price of precious metals. Most of this work is by small companies or lessees.

The principal production in this area has been derived from the turquoise mines, active from 1894 to 1906. The value of the turquoise produced is not known, but undoubtedly it was considerable. Incomplete statistics of the metal production from 1906 to 1932 give 432 tons of ore containing chiefly gold, with small amounts of silver, lead, and copper, valued at $7,655, or an average of $17.72 per ton. Small amounts of feldspar, fluorspar, and mica also have been produced.

Nippeno Group

The Nippeno group of five patented claims 2 miles east of Crescent is owned by J. W. Wilson, of Las Vegas, Nev., and Harry Trehearne, of Nipton, Calif. From 1905 to 1907 this property was worked by the Nippeno Mining Co. On February 1, 1937, the property was leased to George Miller and associates. Production from the Nippeno property is said to have been about $40,000, mostly in shipping ore.

Some years ago a mill was built near Crossman's Wells at Crescent to treat the ore. This mill is now owned by Lester A. Scott of Los Angeles. The mill is equipped with a Huntington No. 2 crusher, 5 stamps (each weighing 1,050 pounds), an amalgamation plate, and a small service concentrating table. Power is furnished by a 6-cylinder gasoline engine. Capacity of the mill is about 15 tons per day. Judging from the tailings pile, about 200 tons of ore was treated.

Development work consists of a shaft inclined 50° and 535 feet deep, with levels at 100-foot intervals, and two shafts each about 100 feet deep. Total workings are about 3,000 feet. Mining equipment comprises a 12-horsepower Fairbanks-Morse geared hoist, a Rix compressor (size 6, 8 inches) direct-connected to a gasoline engine, mining tools, blacksmith shop, and camp buildings.

The formation is granite and diorite. The ore occurs in three narrow veins ranging in width from a few inches to 2 feet and averaging about 10 inches. Values are chiefly in gold with small amounts of silver.

Budget Group

The Budget group of four unpatented claims owned by F. B. McCaughey, of Las Vegas, Nev., and associates is about 1 1/2 miles south of Crescent Peak and 9 miles east of Nipton, Calif. This property was discovered about 1906, at which time a small amount of development work was done, and as nothing of importance was found it was abandoned. It was relocated by the present owners in 1936.

Development work consists of several shallow shafts and several adits, comprising in all about 600 feet in workings. Mining equipment comprises a Ford engine compressor, jackhammer, mining tools, and blacksmith shop. The only production from the property has been 200 tons of ore, averaging about $50 per ton, shipped by the present owners. A shipment made to the American Smelting and Refining Co. from this property on April 21, 1936, furnished the following data:

—Metal quotations: Au	$34.9125				
	Ag	.77			
	Pb				
	4.60 less 3.525				
		Ounce			
Settlement assay:	Au	1.3875			
	Ag	1.95			
		Percent			
	Pb	3.6			
	Cu	.1			
	Insol.	81.6			
	S	.5			
	Fe	7.8			
Value per ton:	Au at 31.81825		$44.15		
	Ag 95 percent less 0.5 oz. at 0.77		1.12		
	Pb less 50 percent at 0.215		.39		
	Gross value		45.66		
Smelter charge:					
	Base charge		$3.50		
	10 percent excess over $20.00 gr. max		2.00	5.50	
	Net value per ton			40.16	
	Pounds				
Gross weight	25,540				
Weight, 225 sacks	255				
Wet weight	25,285				
Less 1.7 percent moisture	429				
Dry weight	24,856 or 12.428 tons at $40.16				$499.11
Deductions:	Freight $4.50 per ton	$57.47			
	Emergency freight 7 percent	4.02			
	Switching	1.35			
	Emergency freight 10 percent	.14			
	Total	62.98			62.98
	Net proceeds				436.13

The ore occurs in a quartz vein striking about north 25 east, dipping 35° to 50°, and ranging from a few inches to 2 feet in width, averaging about 8 inches. The formation is gneiss. Values are chiefly in gold, with small amounts of silver and lead. The quartz is stained with iron and manganese oxides. Wulfenite and a vanadium mineral, probably vanadinite, occur in the vein filling.

Red Star Group

The Red Star group of five unpatented claims, owned by Harry Trehearne of Nipton, Calif., lies 4 miles north of Crescent. In February 1937 the property was under bond and lease to Ralph Wyman of Glendale, Calif., and his associates. There is no record of any production from the property.

Development work comprises a shaft about 100 feet deep and a crosscut 60 feet long driven from the bottom of the shaft. This work was done in the early days. When the writer visited the property a crew of 3 men was installing equipment in preparation for sinking a new shaft 200 feet. The equipment included a 1-drill Worthington compressor, an air-operated hoist, and mining tools.

The general country rock is gneiss. The so-called "vein" appears to be a mass of quartzite carrying gold and silver values. The deposit is reported to have a width of 25 feet and to average $7 to $8 per ton at current metal prices.

Lily Group

The Lily (formerly known as the Calavada) group of five unpatented claims owned by Charles Johnson, of Nipton, Calif., is about 5 miles a little south of east from Nipton. The claims lie on the Nevada-California boundary line, and most of the ground lies in San Bernardino County, Calif. When the writer visited the property in February 1937, it was under bond and lease to J. F. McClish of Glendale, Calif. There is no record of any production. Development work consists of one shaft 150 feet deep, 2 shafts each 60 feet deep, several short adits, and scattered surface workings, altogether totaling about 500 feet. In February 1937 the property was idle.

F. L. Ransome[9], who visited the property in 1906, comments as follows:

> About 3 miles southwest of Crescent, on the other side of a gneiss ridge, is the Calavada property. Here, also, the deposit appears to be a much shattered and silicified bed or beds of quartzite resting on the gneiss and dipping gently south at an angle very little steeper than the general slope of the ridge upon which the deposit lies. The quartzite mass is fully 1,000-feet wide from east to west, and the distance from the highest point on the slope to the point where the deposit passes beneath the alluvium of the desert is probably still greater. A shaft 40 feet deep has been sunk at one place without reaching the underlying gneiss.
>
> The material is much shattered and in many places is a breccia of quartz fragments. It is generally rusty and varies in hardness. The deposit has been carefully and conservatively sampled over an area about 1,000 feet square for the purpose of determining its availability for extensive cyanide treatment, the proposition being to handle the material with steam shovels. These assays rarely fall below $2 per ton, and some of them run as high as $12. The average for the whole area is approximately $3 per ton. The deposit is separated from the railway by a mile of gently sloping plain, and water can be pumped from wells sunk in a playa or dry lake less than 3 miles away.

Double Standard Group

The Double Standard property, comprising two patented claims controlled by the Mitchell Investment Co. of Denver, Colo., is 7 1/2 miles by road northeast of Nipton, Calif. At the time of the writer's visit the property was under lease to W. C. Kelsey, of Searchlight Nev., who was subleasing it to F. F. Galbraith and partner. Royalty paid by Galbraith is 25 percent on the net returns. In 1905 this property produced several thous-nd dollars in shipping ore. In 1936 Galbraith and partner mined nearly 200 tons of ore valued at from $25 to $65 per ton.

Development work comprises four shafts, the deepest of which is 100 feet, and several short tunnels, totaling in all about 500 feet. Mining is done by hand methods,

The property lies on a slope of a steep mountain, and the lessees pack the ore 1 mile off the mountain on burros. with a pack team of six burros an average of 3 tons of ore per day can be hauled, each burro carrying two sacks of ore weighing about 110 pounds each. Cost of feeding the burros averages $45 per month. The ore is hauled to Searchlight, a distance of 22 miles, and treated in the Kelsey custom mill. Trucking charge to Searchlight is $3 per ton on contract. Milling charge is $4 per ton, and payment is made for 90 percent of the assay value of the gold figured at $33 per ounce and 90 percent of the assay value of the silver less one ounce at 70 cents per ounce.

The quartz veins dip about 60° and range in width from 2 inches to 3 feet, averaging about 1 1/2 feet. Formation is granodiorite. Values are in gold and silver associated with small amounts of lead and copper.

Colonel Sellers Claim

The Colonel Sellers property, 3 miles north of Crescent, comprises one patented claim owned by the Mitchell Investment Co. of Denver, Colo. In 1937 it was under lease to Dewey Yeager of Crescent, Nev. Royalty payment is 10 percent of the net mill or smelter returns. This property was first worked about 1905, but there is no record of any production. In the latter part of 1936, Yeager shipped one carload of ore that averaged $20.80 per ton.

Equipment includes a 1-drill giant compressor belt-driven by an automobile engine, jackhammer, and mining tools. Property has been explored by several adits and scattered surface workings, totaling in all about 500 feet.

The ore occurs in narrow quartz veins that dip about 60°. Values are chiefly in gold associated with minor amounts of silver, copper, and lead. Formation is gneiss.

Cumberland Group

The Cumberland (formerly known as the Oro Fino) group of five unpatented claims owned by Oscar C. Scott of Searchlight, Nev., lies 1 1/2 miles east of Crescent. Production has been about $1,000 in shipping ore. When the writer visited the property it was inactive. Development work consists of a shaft 137 feet deep, inclined 30°, and subsidiary workings, totaling about 350 feet. There is no equipment on the property.

Formation is gneiss and granite. The ore occurs in narrow quartz veins and the values are chiefly in gold associated with silver, lead, and copper. Specimens of ore found on the dump showed vanadinite and wulfenite.

Turquoise Deposits

Turquoise occurs in two areas in the Crescent district. One deposit, known as the Morgan mine, is 1 1/2 miles south of Crescent, and two other deposits, known as the Smithson-Phillips and the Toltec mines, are on the south slope of Crescent Peak about 3 1/2 miles a little south of east from Crescent. These deposits are of interest in that they were worked in ancient times by Indians, as is evidenced by stone hammers, stone slabs, and other crude tools used to mine and polish the turquoise that were found in the vicinity. In 1896 the Crescent Peak deposits were rediscovered by George Simmons. In 1896 they were acquired by J. R. Wood of New York City, and in the following year operations began under the name of Toltec Gem Mining Co. This company worked the deposits up to 1906 and, judging from the extent of the old workings, a large amount of turquoise was produced. In 1906 a turquoise weighing 320 carats and valued at $2,600 was found in one of the mines.

The Crescent Peak deposits are covered by six patented claims now owned by the county for nonpayment of taxes. In recent years the only activity has been by lessees who confined their activities to screening the dumps. The road to these deposits has been washed out. and the last 1 1/2 miles up the mountain must be traversed on foot.

Mining activity has been confined to two areas about 1 1/2 miles apart. Total workings consisting of a shaft about 240 feet deep, a number of short adits, and open cuts, all of which total about 1,000 feet.

The formation in the vicinity of the turquoise area is a decomposed rock of granitic composition. The rock has been extensively fractured, and in these fractures veinlets of quartz and turquoise have been formed. Some of the fractures are filled with clay gouge in which nodules of turquoise occur coated with kaolin, which has to be chipped off before the quality of the gem can be determined. The color ranges from light to dark blue, texture is dense, and virtually free from matrix. Fragments of turquoise occur in the dumps, and one fragment found by the writer was 1 inch thick and approximately 1 inch square. Most likely most of the material in the dump was rejected because of its light color.

The Morgan mine, covered by one unpatented claim, is owned by Charles Johnson of Nipton, Calif. It was last worked by G. W. Morgan about 1906. Production is said to have been several hundred pounds of fine quality gem turquoise. Development consists of an adit 150 feet long and several open cuts totaling about 200 feet of workings.

Here the turquoise occurs in narrow seams and veinlets in monzonite cut by rhyolite. Fragments of pale-blue to deep-blue turquoise were observed on the dumps. These dumps have been pretty well picked over by Indians.

Feldspar Deposit

A large deposit of feldspar occurs in a pegmatite dike on the west slope of Crescent Peak 9 miles east of Nipton, Calif., the nearest shipping point. The deposit is said to be owned by Thomas Thorkildson of Los Angeles, Calif. Shipments of feldspar were made from the property about 1923. Judging from the excavations, probably about 500 tons of feldspar were shipped. In recent years the property has been inactive.

The feldspar was quarried from several open-cuts, the largest of which is about 40 feet long, 18 feet wide, and an average of 15 feet high. As seen in the open-cuts, the pegmatite consists of about 75 percent feldspar, 20 percent quartz, and 5 percent mica. By selective mining the feldspar can be obtained free from the associated minerals.

Fluorspar Deposits

Small deposits of fluorspar occur in a low range of hills 4 miles due east from Nipton, Calif. Deposits were discovered by Charles Johnson of Nipton in 1914. In 1925 a small amount of development work was done by Thomas Thorkildson of Los Angeles, Calif., but only a few tons of fluorspar have been shipped. The deposits were relocated by Allen Stolte of Searchlight, Nev., in 1934, and presumably he is the present owner. Development work consists of a short tunnel and several open-cuts totaling in all not more than 50 feet. About 6 tons of sorted fluorspar is on the dumps.

The fluorspar occurs in narrow veins ranging in width from several inches up to 15 inches and dipping about 60°. Formation is gneiss. In places the fluorite is stained with copper. The veins so far discovered are too small to produce any appreciable amount of fluorspar.

Mica Deposit

In 1935 about 5 tons of scrap mica was produced from a pegmatite dike 2 miles southeast of

Crescent along the side of the old stage road between Searchlight, Nev., and Ivanpah, Calif. The deposit is owned by H. M. Benson of Crescent, Nev. The only work done on it is an open cut about 10 feet deep and 10 feet wide.

The pegmatite dike is 15 feet wide and consists of an intergrowth of quartz, feldspar, and mica. The mica crystals are small and the material is suitable only for scrap purposes. One-quarter mile north of the pegmatite dike is a prominent outcrop of solid white quartz at least 20 feet wide. The formation is granite.

ELDORADO (COLORADO) DISTRICT
(Gold, Silver)

The Eldorado, also known as the Colorado, district comprises an area 12 miles long and 6 miles wide in the north end of the Opal Mountains in southeastern Clark County. It includes the mines in the vicinity of Knob Hill on the west and those in the Capitol area on the east. The camp of Nelson in the central part of the district is 25 miles by road south of Boulder City, Nev. This district is one of the oldest in the State and mining has been done here almost continuously for nearly 80 years. It was organized as the Colorado district in 1861. Old arrastres and prospect holes reported to have been found in this area in the sixties indicate that mining was carried on by Spanish adventurers probably several hundred years ago. The first authentic discovery was in 1857 on the Honest Miner claim, now a part of the Eldorado Rand property. The Techatticup (an Indian word said to signify either "white flower" or "plenty for all") mine was purchased from the original locaters in 1862 by the Southwestern Mining Co. The mine was opened in 1863, and in the following year a 15-stamp chlorination mill was erected at the mouth of Eldorado Canyon on the Colorado River. The salt required for the treatment of ore was mined from the Virgin River salt beds near St. Thomas, Nev., and transported by boat down the Virgin and Colorado Rivers.

The Southwestern Mining Co. dominated the district until about 1897 as it owned virtually all of the productive mines, including the Techatticup, Wall Street, Quaker City, Mocking Bird, Jubilee, Honest Miner, Morning Star, Red Butte, Rover, and others. The company was at first controlled by the Barker brothers and later by Joseph Wharton, a prominent figure in the steel industry of Pennsylvania. During its greatest activity the company owned several mills and operated its own fleet of river boats for bringing in supplies from the Pacific coast. After 1897 the properties were either sold or leased to a number of small companies, which operated with varying success. Speaking of the Eldorado district in 1906, Ransome[10] states:

> It is a little surprising that a district once alive with activity should have attracted so little outside notice. This, however, is partly accounted for by the overshadowing predominance of the Comstock, Eureka, Ely, and other districts noted in the early history of mining in Nevada and by the isolation of Eldorado Canyon. The rich ore shipped from the canyon in early days was taken down the Colorado by boat to Needles, Yuma, or the Gulf of California.

The production of the Eldorado district from 1907 to 1935 is shown in table 2. The bulk of the production, however, was made from 1864 to 1900, and for this period no accurate data are available. Yeoman Briggs, of Nelson, Nev., has compiled production statistics on all the individual properties in this area for the Nelson-Searchlight power district and, according to Briggs, figures the total production has been about $10,000,000. Briggs obtained his data from many sources, and due allowances must be made for old production figures, which have a way of increasing with time.

TABLE 2. - Gold, silver, copper, lead, and zinc production from Eldorado Canyon District, Clark County, Nevada, 1907-1935

(Compiled by Charles White Merrill, supervising engineer, San Francisco office, Mineral Production and Economics Division, U. S. Bureau of Mines)

Year	Placer						Lode			
	No. of mines	Gold		Silver		Total value	No. of mines	Ore, short tons	Gold	
		Fine ounces	Value	Fine ounces	Value				Fine ounces	Value
1907	----	----	----	----	----	----	5	417	306.94	$ 6,345
1908	----	----	----	----	----	----	1	25	36.28	750
1909	----	----	----	----	----	----	3	48	74.01	1,530
1910	----	----	----	----	----	----	1	431	139.08	2,875
1911	----	----	----	----	----	----	6	541	125.53	2,595
1912	----	----	----	----	----	----	5	137	37.20	769
1913	----	----	----	----	----	----	4	276	125.29	2,590
1914	----	----	----	----	----	----	2	162	81.56	1,686
1915	----	----	----	----	----	----	3	6,547	3,047.58	62,999
1916	----	----	----	----	----	----	9	24,868	10,310.26	213,132
1917	----	----	----	----	----	----	7	12,689	6,631.18	137,079
1918	----	----	----	----	----	----	4	15,302	5,424.92	112,143
1919	----	----	----	----	----	----	4	13,875	3,553.63	73,460
1920	----	----	----	----	----	----	3	9,160	2,280.01	47,132
1921	----	----	----	----	----	----	1	2	1.06	22
1922	----	----	----	----	----	----	----	----	----	----
1923	----	----	----	----	----	----	1	160	80.83	1,671
1924	----	----	----	----	----	----	1	917	1,001.46	20,702
1925	----	----	----	----	----	----	5	708	539.14	11,145
1926	----	----	----	----	----	----	4	2,104	1,380.57	28,539
1927	----	----	----	----	----	----	3	1,155	945.05	19,536
1928	----	----	----	----	----	----	3	1,183	468.56	9,686
1929	----	----	----	----	----	----	3	2,207	1,434.46	29,653
1930	----	----	----	----	----	----	2	3,350	1,277.05	26,399
1931	----	----	----	----	----	----	1	3,689	2,174.55	44,952
1932	2	110.68	$2,288	27	$ 8	$2,296	8	3,463	968.34	20,017
1933	3	40.32	1,031	11	4	1,035	7	1,823	663.20	16,951
1934	1	3.19	111	----	----	111	3	4,086	927.55	32,418
1935	3	11.41	399	1	1	400	11	3,565	856.89	29,991
Total	----	165.60	3,829	39	13	3,842	----	112,890	44,892.18	956,767

5883

- 28 -

TABLE 2. - Gold, silver, copper, lead, and zinc production from Eldorado Canyon District, Clark County, Nevada, 1907-1935 (Continued)

(Compiled by Charles White Merrill, supervising engineer, San Francisco office, Mineral Production and Economics Division, U. S. Bureau of Mines)

Year	Lode									Average recoverable value of ore per ton[1/]	Total value (lode & placer)
	Silver		Copper		Lead		Zinc		Total value		
	Fine ounces	Value	Pounds	Value	Pounds	Value	Pounds	Value			
1907	4,533	$ 2,992	—	—	—	—	—	—	$ 9,337	$22.39	$ 9,337
1908	—	—	—	—	—	—	—	—	750	30.00	750
1909	3,212	1,670	—	—	2,232	$ 96	—	—	3,296	68.67	3,296
1910	2,335	1,261	—	—	—	—	—	—	4,136	9.60	4,136
1911	3,799	2,013	—	—	32	1	—	—	4,609	8.52	4,609
1912	10,379	6,383	157	$ 26	442	20	—	—	7,198	52.54	7,198
1913	346	209	297	46	—	—	—	—	2,845	10.31	2,845
1914	1,650	912	—	—	292	11	—	—	2,609	16.10	2,609
1915	56,859	28,828	—	—	180	8	—	—	91,835	14.03	91,835
1916	241,789	159,097	1,205	296	11,771	812	—	—	373,337	15.01	373,337
1917	90,618	74,669	281	77	4,956	426	—	—	212,251	16.73	212,251
1918	103,672	103,672	119	29	3,430	244	—	—	216,088	14.12	216,088
1919	113,039	126,604	—	—	—	—	—	—	200,064	14.42	200,064
1920	85,113	92,773	—	—	148	12	—	—	139,917	15.27	139,917
1921	22	22	—	—	—	—	—	—	44	22.00	44
1922	—	—	—	—	—	—	—	—	—	—	—
1923	2,283	1,872	—	—	—	—	—	—	3,543	22.14	3,543
1924	2,265	1,518	1,189	156	7,870	630	—	—	23,006	25.09	23,006
1925	3,621	2,513	736	105	6,674	581	—	—	14,344	20.26	14,344
1926	17,714	11,054	2,816	394	18,808	1,505	1,800	$ 135	41,627	19.78	41,627
1927	2,912	1,651	927	121	4,435	279	—	—	21,587	18.69	21,587
1928	1,000	585	568	82	2,242	130	—	—	10,483	8.86	10,483
1929	9,813	5,230	264	46	3,860	243	—	—	35,172	15.94	35,172
1930	5,507	2,120	432	56	1,537	77	—	—	28,652	8.55	28,652
1931	6,436	1,866	—	—	—	—	—	—	46,818	12.69	46,818
1932	3,690	1,041	—	—	499	15	—	—	21,073	6.09	23,369
1933	2,796	979	250	16	5,664	210	—	—	18,156	9.96	19,191
1934	6,731	4,348	1,813	145	9,911	367	—	—	37,278	9.12	37,389
1935	7,359	5,289	443	37	2,271	91	—	—	35,408	9.93	35,808
Total	789,493	641,171	11,497	1,632	87,254	5,758	1,800	135	1,605,463	14.22	1,609,305

1/ Not to be confused with average assay value.

Eldorado Canyon Development Co

The Eldorado Canyon Development Co., D. P. Simons of Nelson, Nev., president, controls the Wall Street group (three patented claims), the Mocking Bird (one patented claim), and seven unpatented claims. The Wall Street mine in the early days is said to have produced $1,750,000, and the Mocking Bird claim $240,000. In 1935 the present company erected a 50-ton daily capacity amalgamation-flotation mill on the property. Since November 1935, When the mill was completed, up to February 1937, 6,500 tons of ore were treated, from which bullion and concentrates valued at $55,000 were produced. Most of the ore was obtained from the dumps and stope fills. An average of 12 men are employed.

Development work comprises a number of shafts, the deepest of which is about 200 feet, and approximately 1 mile of underground workings. Most of the development work has been done on the Wall Street group of claims.

The Wall Street vein strikes about east and west, with a dip of 15° to 30° to the south. The old stopes mined through to the surface show widths up to 15 feet. The ore occurs in a shattered zone of monzonite in a gangue of quartz, calcite, and country rock. Values are in gold and silver.
to a 75-kva generator. The cost of power is 1 1/2 cents per kilowatt hour. Water for milling is pumped from a shaft in the vicinity.

Diamond Gold Mining Co.

The Diamond Gold Mining Co., P. A. Simon, Las Vegas, Nev., president and general manager, controls 12 patented mining claims and 1 patented millsite, the latter at the mouth of the Eldorado Canyon on the Colorado River, 6 miles east of the mine. The property is worked under bond and lease from the Joseph Wharton estate, administered by the Girard Trust Co. of Philadelphia, Pa. It included the Techatticup mine, Which has been the largest producer in the Eldorado district.

Production of the Techatticup mine, most of which was made prior to 1900, is estimated at about $3,500,000. Judging from the extent of the mine workings, the tailings pile (about 150,000 tons) near the present mill site, and the ore that was treated in the early day's in mills on the Colorado River, this estimate is probably fairly accurate. From 1913 to 1920 the mine was operated by Martin Heller of New York City. According to P. A. Simon, production for this period was 60,000 tons of ore averaging 0.55 oz.gold and 9.5 oz.silver, having a value of $17.82 per ton. This ore was treated in the cyanide mill formerly on the property. From 1923 to 1930 the Stile brothers, lessees, produced $48,000 in crude ore and flotation concentrates. Under the management of P. A. Simon, the mine has been operated intermittently for the past four years. In February. 1937 the Diamond Gold Mining Co. employed an average of 35 men.

Development work comprises about 3 miles of underground workings. The Techatticup inclined shaft has a depth of 600 feet, with levels at about 100 foot intervals. An adit 180 feet long connects with the Techatticup shaft about 200 feet below the outcrop of the vein. An interesting feature of the property is that most of the development work has been in ore.

Mining equipment includes three compressors an Ingersoll-Rand (12 by 12 inches), a Union (10 by 12 inches). and an Ingersoll-Rand Imperial type 10 (10 by 16 by 12 inches) drill sharpener, rock drills and other mining tools. Hoisting is done in a 1-ton skip operated by 20-horsepower Fairbanks-Morse geared hoist. There is a 50-ton daily capacity amalgamationflotation mill on the property. Power for mining and milling is furnished by a 200 horsepower, Fairbanks-Morse, Y-type, Diesel engine.

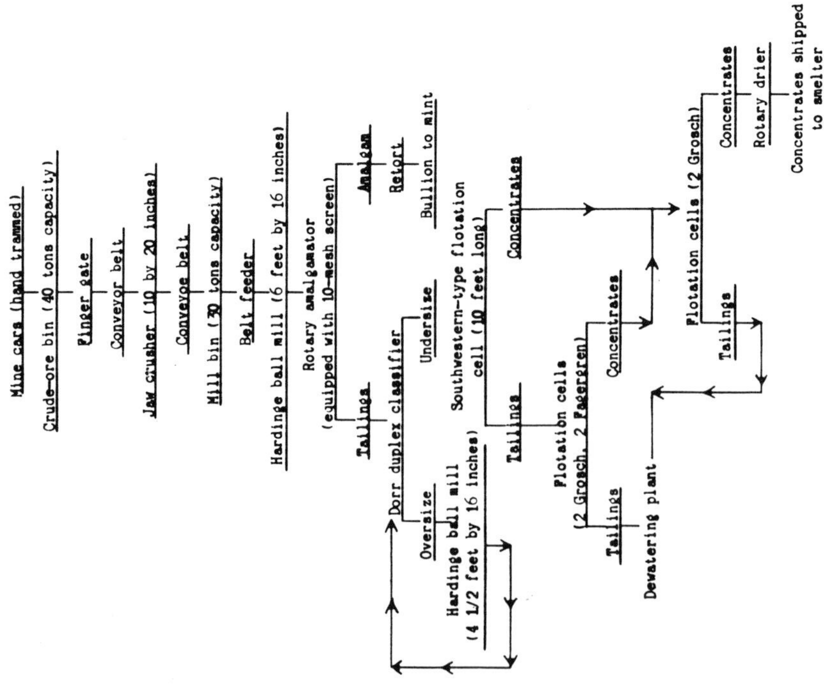

Figure 5.- Flow sheet of the Diamond Gold Mining Co. mill, Eldorado district, Clark County, Nev.

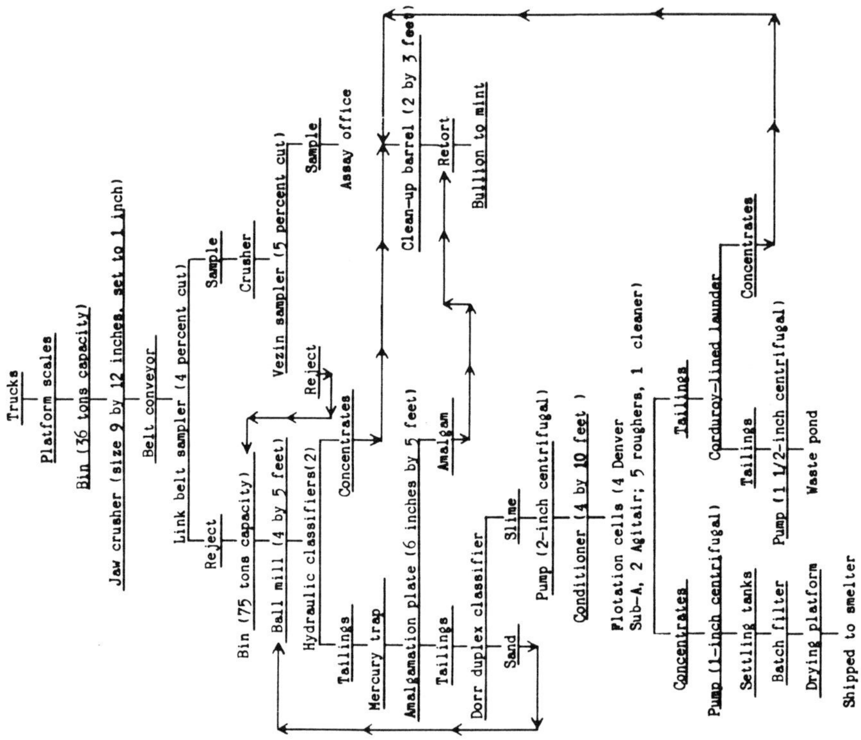

Figure 6.- Flow sheet of Kinney custom mill, Goodsprings, Clark County, Nev.

There are two veins on the property the Savage and the Techatticup. In the former silver values predominate, while in the latter values are chiefly in gold. The ore occurs as a network of stringers in monzonite. The strike of the veins is about east and west, and the dip varies from 20° to 70° to the north. The width varies from 2 to 12 feet, averaging about 5 feet. The Techatticup vein is traceable on the surface for several thousand feet. The gangue minerals are principally quartz, calcite, and shattered country rock, with subordinate amounts of pyrite, sphalerite, galena, and chalcopyrite. A little below the 400 foot level the veins are displaced by a flat fault. The monzonite stands well, and the open stope method of mining is employed.

The flow sheet of the mill is shown in figure 5. Water for milling is hauled from the Colorado River, 6 miles distant, in tanks mounted on trucks, at a cost of $2 per 1,000 gallons. In February 1937, to eliminate the water haul, the company was installing a 4inch pipe line 5.7 miles long from the Colorado River to the mill. The total lift will be 1,842 feet, and water will be pumped to the mine in two stages by triplex pumps. The value of the ore treated is from $12 to $18 per ton, averaging about $15. Recovery by amalgamation and flotation averages 85 percent. Flotation concentrates have a value of from $500 to $1,400 per ton.

Eldorado Rover Mining Co.

The Eldorado Rover Mining Co., P. A. Keegel, of Las Vegas, Nev., president and general manager, controls the Rover, Flagstaff (Duncan), Rambler, and Sky Lark groups, comprising 17 patented claims. Production from these groups of claims has been about $550,000 of which $360,000 was produced from the Flagstaff. In February 1937, 12 men were employed and the company was preparing to operate its 75 ton daily capacity cyanide mill.

Development comprises three inclined shafts, the deepest of which is the Flagstaff, which is inclined 40o and is 500 feet deep. Underground workings total about 4 miles. Mining equipment includes a 25-horsepower Fairbanks-Morse geared hoist of the Duncan shaft, a 15horsepower Fairbanks- Morse geared hoist, and an Ingersoll Rand Imperial type 10, 2-stage compressor belt driven by a 60-horsepower Fairbanks-Morse distillate engine at the Rambler shaft. The mill is equipped with a Blake type crusher (10 by 16 inches), belt driven by a 25-horsepower Fairbanks-Morse distillate engine; 10 stamps (1,250 pounds each); a pebble mill (16 feet long and 4 1/2 feet in diameter); a Dorr duplex classifier, Wilfley table, two agitators (18 feet by 12 feet), a thickener (24 by 12 feet). and Oliver filter (9 feet in diameter), solution tanks, and other cyaniding equipment. Power is furnished by a 90-horsepower, Fairbanks-Morse, Y-type, 3-cylinder, Diesel engine and a 40-horsepower, Fairbanks-Morse, distillate engine. Water for milling is pumped from the Duncan and Rambler shafts.

The three veins on the property are the Rambler, Rover, and Flagstaff, which occur either in monzonite or at the contact of monzonite and andesite. In February 1937 mining was confined to the Rover vein. This vein strikes east and west, dips about 70°, and ranges in width from 2 to 6 feet. It outcrops for a distance of several thousand feet. The ore occurs in a gangue of shattered country rock with bunches and stringers of quartz and calcite. Values are chiefly in gold, with minor amounts of silver.

Eldorado Rand Mine

The Eldorado Rand mine comprises 27 claims (three of them patented) controlled by the Guaranty Liquidating Corporation of Los Angeles, Calif. Recently this property was leased for 10 years to an eastern syndicate represented by John Daam of San Francisco, Calif. When the writer visited the property in February 1937 the syndicate was planning to develop the mine further and to construct a mill on the Colorado River, about 8 miles east. This property has been a consistent producer, and past production, chiefly from the Quaker City and Rand patented claims, has been approximately $1,000,000.

Development comprises about 6,000 feet of underground workings. The Rand shaft is 350 feet deep. Equipment includes a 40-horsepower Fairbanks Morse geared hoist, a compressor (10 by 10 inches) belt driven by a Fairbanks-Morse distillate engine, and a 10-stamp cyanide mill originally built in 1907. The mill equipment is in poor condition. Adjacent to the mill is a tailings pile estimated to contain 50,000 tons, but the writer could not learn whether the tailings contained sufficient values to be worth retreating.

The ore occurs in a series of veins varying from 3 to 6 feet in width in a monzonite formation. The veins are distinguished by their diversity in dip, varying from nearly horizontal to 80o. Values, chiefly gold and silver, occur in fissure zones in a gangue of country rock, calcite, quartz, pyrite, galena, and sphalerite.

Crown Queen Group

The Crown Queen group of three patented claims, known as the Eldorado Crown mining Co., is controlled by Martin Heller, 120 Broadway, New York City. In February 1937 the property was worked by R. T. Roe, W. R. Cozart, and Osborne Gentry under 1year lease with a royalty of 10 percent of net mill or smelter returns. The lessees were working at the surface on the east end of the vein, and several hundred tons of ore, said to average $15 per ton, had been exposed. Arrangements were being made to treat the ore at the Eldorado Rover mill, 2 miles distant. Prior to 1907 the property is said to have produced $450,000 in milling ore treated at the Techatticup mill.

Development work comprises a shaft 300 feet deep and inclined 52° and more than a mile of underground workings. Equipment includes a Rix 1 drill compressor belt driven by 6-cylinder automobile engine.

The main vein strikes east and west and dips 50° to the south. Formation is monzonite and andesite. Judging from the old stopes mined through to the surface, the average width of the vein is about 5 feet. Gold and silver values occur in a network of small stringers in a fairly well defined zone in shattered country rock.

Eldorado Empire Mining Co.

The Eldorado Empire Mining Co., C. E. L. Gresh, of Nelson, Nev., president and principal owner, controls the Empire group of three claims (one patented) adjoining the Wall Street group on the east. Although operations have been carried on intermittently for a number of years, very little has been produced and in February 1937 it was inactive.

Development work comprises a shaft 200 feet deep and. other workings totaling in all about 2,000 feet. Mining equipment includes a 6horsepower Fairbanks-Morse geared hoist, a 1drill Ingersoll-Rand compressor, assay office, and blacksmith shop.

Equipment for a partly completed mill consists of . small Wheeling crusher, a Marathon rod mill, and a 15horsepower Fairbanks-Morse distillate engine.

According to Gresh, the ore occurs in a brecciated zone 40 feet wide, in monzonite. When the writer visited the property, about 50 tons of ore on the dumps were reported to average about $12 per ton in gold and silver. It was planned to treat this ore in the nearby Eldorado Canyon Development Co. mill.

Golden Empire Group

The Golden Empire group of eight claims owned by John S. Sartain is in the western part of the Eldorado district in the vicinity of Knob Hill. This property has been worked intermittently for many years and production is reported to have been about $50,000 in shipping ore. In February 1937 the property was under lease to R. W. Cate, with royalty payment of 10 per cent on the net smelter returns. In 1936 Cate shipped 50 tons of ore averaging $50 per ton. The ore is mined by hand methods. The haul to Boulder City, 32 miles from the property, costs $3 per ton. Development work comprises four shafts, each about 100 feet deep, and subsidiary workings, totaling in all about 1,000 feet.

The ore occurs in narrow veins from a few inches to 2 1/2 feet wide, striking east and west and dipping 60° to 80° to the south. Values are in gold and silver. Formation is gneiss and schist.

Other Properties

The Rich Hill group of two claims owned by J. Taylor, the Silver Legion group of nine claims owned by John S. Sartain and associates, and the Combi- nation group of two claims are in the Knob Hill section of the Eldorado district. When the writer visited here in February 1937 these properties were inactive.

In general, the ore in the Knob Hill area is in east and west quartz veins in gneiss and schist. Some of the veins, although narrow, are very persistent and can be traced for several thousand feet. The values are in gold and silver associated with small amounts of pyrite, galena, and sphalerite.

In the Capitol area, in the eastern part of the district, a number of small properties were operated in former years. The first discovery in this area was made in 1894 by John Appel. The principal property is the Capitol group of five patented claims owned by Thomas F. Murphy of Los Angeles, Calif. Production is reported as $100,000.

Other properties are the Briggs Capitol group of 17 patented claims owned by Yeoman Briggs of Nelson, Nev., the Wallace group of 10 patented claims owned by William C. Wallace of Oakland, Calif., and associates, the Nevada Eagle group of six patented claims owned by the J. T. Weyerhaeuser interests of Tacoma, Wash., and various other groups of unpatented claims.

In the Capitol area the ore occurs as numerous narrow quartz veins in monzonite and andesite. Values are chiefly in gold, with small amounts of silver. Nearly all the properties in this section are in the prospect stage of development.

GOLD BUTTE DISTRICT
(Gold, Zinc, Copper, Mica, Magnesite)

The Gold Butte district is in southeastern Clark County in the south ern end of the Virgin Range. It includes the territory south of Gold Butte lying between the Nevada Arizona boundary line on the east and the Virgin River on the west. mining was begun in this area in the eighties. A small boom occurred in 1908, when the camp of Gold Butte was established, and although a number of small companies were organized to work various properties, no important discoveries were made. The total production from the district has been about $75,000, mostly in shipping ores. In 1936 the principal operation in the district was that of the Lake Shore Mining Co.

The geology of the Gold Butte mining district has been described by Hill[11].

The ore deposits are of two types replacement deposits in limestone and quartz veins in gneiss and granite. The ore in the limestone consists of oxidized copper, lead, and zinc minerals. The values in the quartz veins are chiefly in gold.

A small amount of sheet mica has been produced from pegmatite dikes in the vicinity of Gold Butte.

Lake Shore Mining Co.

The property of the Lake Shore mining Co., comprising the Utah group of four unpatented claims owned by O. W. Yates, A. W. Lawson, and Fred Gibson of Las Vegas, Nev., is located about 15 miles south of Gold Butte and 5 miles from the shore of Lake Mead. In 1934 and up to July 1935 the Utah group and other claims were worked by the Gold Cross mining Co., controlled by Salt Lake City interests. The Gold Cross Mining Co. erected a small amalgamating mill on the shore of the Colorado River and treated about 400 tons of ore. This company also shipped 340 tons of ore, averaging $51 per ton, to Utah smelters. The Lake Shore Mining Co. ac quired the Utah group of claims under bond and lease from C. C. McDonald, of Overton, Nev., in 1935 and started operations in September of the same year. Up to February 1937 total production by the present owners was 1,800 tons of shipping ore with a net value of $30,000. Five men are employed. Development work comprises an incline 200 feet deep and other workings, totaling about 400 feet.

The ore occurs in a quartz vein that dips about 8o and follows the hangingwall side of a diabase dike. The width of the vein varies from 6 inches to 6 feet, averaging about 2 feet. The vein and dike are in granite formation.

Due to the flat dip of the vein, most of the ore shipped by the present owners is mined by the opencut method. The granite overburden is drilled with jackhammers and blasted with 40 percent gelatin dynamite and No. 6 caps attached to tape fuse. Compressed air is furnished by a portable compressor. Stripping is done with a lO cubic foot capacity Le Tourneau bulldozer operated by a 60horsepower caterpillar tractor. The average cost of stripping is 12 cents per cubic yard. After the over burden has been removed (a maximum depth of 20 feet), the ore is hand shoveled into a truck and hauled 5 miles to the shore of Lake Mead, where it is loaded onto a barge. The barge is towed by power boat 56 miles to Cashman Docks, and the ore is again shoveled into trucks and hauled to the railroad siding at Boulder City, 6 1/2 miles distant. The barge and power boat are owned by the company. Smelter returns from a shipment of ore to the American Smelting & Refining Co. were as follows:

Metal quotations:	Au	$34.9125 per oz.		
		Oz.		
Settlement assay:	Au	1.5525		
	Ag	.775		
		Percent		
	Insol	88.7		
	S	.4		
	As	.1		
	Fe	4.3		
	CaO	1.5		
Metal payment:	Au at $31.81825 per oz.			$49.40
Treatment charge (base)			$ 3.22	
10 percent excess over $20 (max.)			1.78	
	Total treatment charge		5.00	5.00
		Net value per ton		44.40

	Lb.		
Wet weight	125,740		
Less 2.15 percent moisture	2,704		
Net weight	123,036	or 61.518 tons	
	61.518 tons at $44.40 per ton		$2,731.40
Freight, $6.60 per ton		$414.94	
Emergency freight		25.15	
Demurrage		2.00	
Watching		8.00	
Assaying		6.00	456.09
	Net proceeds		2,275.31

Azure Ridge Group

The Azure Ridge group of four unpatented lode claims owned by John F. Perkins of Overton, Nev., is near the Arizona-Nevada boundary in the southeastern part of the Gold Butte district. The only production from the property was in 1918, when John F. Perkins shipped two carloads of zinc ore and one carload of copper ore. The zinc ore averaged 40 percent zinc and the copper ore 35 percent copper, with small values in gold and silver. This ore was hand-sorted and hauled to St. Thomas for shipment to smelters. Since 1918 the property has been inactive.

This property is in the prospect stage of development and all work has been superficial in character. Development comprises an adit 100 feet long, another of 40 feet, and several shallow shafts, totaling in all about 300 feet of workings. mineralization occurs in a faulted zone in limestone near granite.

Webster Group

A group of seven unpatented claims in Cedar Basin is owned by Mrs. A. G. Webster of Moapa, Nev. In 1937 this property was under option to H. G. Snyder of Salt Lake City, Utah. Development comprises a shaft 130 feet deep and several hundred feet of lateral workings. Property is equipped with a 2-stamp mill (1,050 pounds each), a jaw crusher (6 by 8 inches). and an amalgamation plate (4 feet long and 3 feet wide). Mill equipment is operated by an automobile engine. About 150 tons of ore, averaging 1 ounce of gold per ton, was treated in this mill.

The ore occurs in a quartz vein in the granite. The dip of the vein is about 75o and the average width is 18 inches. Formation is altered granite and schist traversed by pegmatite dikes.

Mica Deposits

In 1873 Daniel Bonelli discovered mica deposits 4 miles east of Gold Butte, but because of their isolation and unfavorable transportation facilities very little mining was done. As far as the writer could learn, the only production has been 5 tons of sheet mica shipped by Bonelli prior to 1900 and 2,500 pounds shipped by Frank Allsop in 1908. In recent years these deposits have not been exploited. William Garrett of Gold Butte is the owner of several unpatented mica claims in this area.

The deposits have been prospected by a number of shallow shafts, the deepest of which is about 40 feet. The mica, associated with garnet, quartz, feldspar, and tourmaline, occurs in pegmatite dikes that cut granitic schists. Due to the manner in which the mica occurs, it is difficult to mine it in large sheets. The mica is said to be of good quality as to transparency, color, cleavage, and flexibility, and the size of the sheets varies from 6 to 15 square inches.

Magnesite Deposit

Magnesite occurs in the vicinity of Horse Springs, 14 miles by road southeast of St. Thomas, Nev., and 9 miles north of Gold Butte. The deposit was located originally by Fay Perkins of Overton, Nev., in 1922. The present owner is Albert Bauer. No production has ever been made. Development comprises several short tunnels and a shaft about 60 feet deep. Fine-grained dolomite and magnesite beds overlain by shale and underlain by limestone and dipping about 30° outcrop for several thousand feet.

GOODSPRINGS (YELLOW PINE) DISTRICT
(Zinc, Lead, Copper, Gold, Silver, Molybdenum, Vanadium, Cobalt, Platinum

The Goodsprings, also known as the Yellow Pine, district covers an area of several hundred square miles in the southern part of the Spring Mountain Range in southwestern Clark County. The town of Goodsprings, the center of mining activities of this area, is 8 miles northwest of Jean, Nev., a station on the Union Pacific R.R.

Potosi mine in the northern part of the district was discovered by a party of Mormons in 1855. Shortly after, a small amount of lead ore was mined from this property and was treated locally in crude smelters in order to obtain lead for bullets. This venture proved unsuccessful and work stopped until 1868. when a group of prospectors found lead ore in the southern part of the area. They located a number of claims and organized the New England mining district, but when they found that the lead carried low silver values the district was abandoned. A member of this party, named Good, camped at the springs in the vicinity after the others had left, and Goodsprings was named for him. After the collapse of the New England district, the only work done for many years was by prospectors from Ivanpah, Calif., and Eldorado, Nev.

The permanent settlement of Goodsprings dates from 1886, when A. G. Campbell, John Denslow, A. E. Thomas, William Smith (all from Utah), and others began a systematic search for lead. They located over 40 claims and did considerable development work. In 1892 a miner by the name of Jonas Taylor, from Ivanpah, Calif., located a cropping of iron oxides and named it the Keystone mine. Samuel Godby of Pioche, Nev., secured from Taylor a short-time option on one-half interest in the Keystone mine for $20,000. Godby hastened to San Francisco, where he sold one-half of the option to a banker for $20,000, thereby securing a one-quarter interest for virtually nothing. The three partners started to work the property and several carloads of ore were shipped to Colorado smelters, one netting them $9,000 and another nearly $16,000. At that time the nearest railroad connection was at Fenner, Calif., on the main line of the Santa Fe, approximately 100 miles to the south. The high-grade ore shipped from the Keystone caused considerable excitement, and at the end of 1892 the camp had a population of 200. In December 1893 a 10-stamp mill was erected to treat Keystone ore at a place in Mesquite Valley then known as Taylors Well but now known as Sandy. The Keystone mine was worked with varying success until 1906, production being about $600,000.

In 1893 a branch railroad was built from the Santa Fe line to a point 45 miles from Goodsprings, and the northern terminus, Manville, became the shipping point. This branch was extended 16 miles nearer Good- springs in 1901. When the San Pedro, Los Angeles, & Salt Lake R.R. (now part of the Union Pacific System) was completed in the spring of 1905, Jean, 8 miles from Goodsprings, became the shipping point. Prior to the completion of this railroad some shipments of lead and copper ore were made over the Santa Fe branch, but the long wagon haul was too expensive for any but the highest-grade ores.

In 1905 the mineral reserves were increased vastly by the recognition of zinc in the ores by C. T. Brown, a mining engineer from Socorro, N. Mex. Prior to Brown's discovery, the earthy material associated with the lead ore (that proved to be hydrozincite) was thrown over the dumps. Following the discovery of zinc in the lead ore, a number of deposits were discovered and shipments of crude ore were made from the Potosi, Hoosier, and other mines in 1905.

In 1906 a large body of lead-zinc ore was opened by Frank Kent on the Rover claim 5 miles west of Goodsprings. This claim had been located and relocated a number of times, and it was believed to be of little value. Following Kent's discovery, this claim became the nucleus of the Yellow Pine mine, the largest producer in the district. The Yellow Pine Mining Co. erected a 75-ton daily capacity zinc-lead concentrator at Goodsprings in 1911 and in the following year built a narrow-gage railroad from Goodsprings to Jean. The period of greatest activity in the district occurred during the World War, when the prices of base metals rose considerably. In 1915 the population of the camp exceeded 1,000, and, in addition to the two major producers, the Yellow Pine and Potosi mines, at least 50 smaller properties were productive.

In 1914 considerable excitement was created by the discovery of platinum in the Boss mine by Harold C. Riddall, a Goodsprings assayer. This discovery caused far more excitement than the facts justified, and a townsite in Mesquite Valley called Platina was laid out by unscrupulous promoters and a number of people were swindled.

In 1930, because of the low prices of the base metals, mining stopped, and the population dwindled to about 15 people. In recent years mining has been confined to the gold and molybdenum deposits. When the writer visited the district in February 1937 the population of the camp was about 150 people.

The production of the district is shown in table 3. In addition to the metals shown in the table, small amounts of platinum, molybdenum, cobalt, and vanadium also have been produced. A feature of the mining development in the district worthy of comment is that virtually all the properties that operated in former years either paid their own way from the start or were financed locally without benefit of outside capital.

No attempt has been made in this report to cover the district in detail. A number of reports have been written on the Goodsprings district and the interested reader is referred to the report by Hewett[12], which is the latest and most detailed.

Zinc-Lead-Silver and Copper

In former years the principal production from the Goodsprings district has been zinc-lead ores derived from a large number of properties. The principal producers have been the Yellow Pine and the Potosi mines. Mining has been confined almost entirely to the shipping of oxidized ores obtained chiefly by selective mining and hand sorting. During the war period, dry concentrating, plants generally employing Stebbins tables, were erected at the Singer, Milford, Fredrickson, Bullion, Anchor, Sultan, Hoosier, and Christmas mines. These dry concentrating plants wore fairly satisfactory, but in general the process was less efficient than wet concentration. In 1914 a roasting Plant was installed at the Potosi mine to treat the zinc ores to effect a saving on freight charges. The zinc-lead deposits are of the replacement type in carbonate rocks. The ratio of lead to zinc is extremely variable, ranging from nearly all zinc to nearly all lead. The zinc is in the form of mixed smithsonite, calamine, and hydrozincite, while the lead occurs as cerussite, galena, and anglesite. Some silver is associated with the zinc-lead minerals. The consensus of opinion among the miners in the district is that the combined price of lead and zinc must be above 14 cents before

TABLE 3. - Gold, silver, copper, lead and zinc production from Yellow Pine (Goodsprings) District, Clark County, Nevada, 1902-1935

(Compiled by Charles White Merrill, supervising engineer, San Francisco office, Mineral Production and Economics Division, U. S. Bureau of Mines)

Year	Lode									
	No. of mines	Ore, short tons	Gold		Silver		Copper		Lead	
			Fine ounces	Value	Fine ounces	Value	Pounds	Value	Pounds	Value
1902	1	2,078	2,136.97	$ 44,175	342	$ 181	----------	----------	----------	----------
1903	4	2,092	2,583.04	53,396	695	375	21,800	$ 2,987	28,000	$ 1,176
1904	1	1,929	2,316.10	47,878	146	85	----------	----------	----------	----------
1905	2	2,394	725.62	15,000	3,707	2,261	----------	----------	290,063	13,633
1906	5	9,481	442.63	9,150	1,573	1,070	67,341	12,996	625,175	35,635
1907	5	4,400	364.46	7,534	2,976	1,964	92,690	18,538	172,800	9,158
1908	11	4,627	376.92	7,792	10,247	5,431	42,144	5,563	720,285	30,252
1909	10	8,664	277.67	5,740	18,461	9,600	392	51	406,353	17,473
1910	16	5,878	58.97	1,219	16,826	9,086	122,925	15,611	1,263,837	55,609
1911	17	8,677	117.27	2,424	47,072	24,948	173,719	21,715	1,617,228	72,775
1912	22	28,386	1.63	34	223,013	137,153	103,398	17,061	6,544,917	294,521
1913	25	29,060	61.32	1,268	192,339	116,173	283,592	43,957	6,204,065	272,979
1914	19	24,537	388.60	8,033	122,703	67,855	156,389	20,800	4,185,208	163,223
1915	35	38,391	40.29	833	100,146	50,774	262,600	45,955	4,620,243	217,151
1916	54	71,155	693.36	14,333	156,492	102,972	494,604	121,673	8,349,850	576,140
1917	79	64,260	515.50	10,656	219,789	181,106	764,733	208,772	9,298,706	799,689
1918	38	40,614	235.52	4,869	140,211	140,211	400,792	98,996	6,459,836	458,648
1919	26	19,273	563.55	11,650	139,656	156,415	130,395	24,253	4,217,739	223,540
1920	24	18,511	158.33	3,273	96,557	105,247	76,567	14,088	3,927,280	314,182
1921	5	287	1.09	23	2,195	2,195	119	15	186,169	8,378
1922	10	1,861	.95	20	6,618	6,618	52	7	316,125	17,388
1923	13	14,758	5.26	109	31,678	25,976	1,855	273	1,472,113	103,048
1924	13	21,383	7.00	145	123,650	82,845	16,267	2,131	7,710,851	616,868
1925	23	20,435	14.39	298	72,045	49,999	6,900	980	5,287,587	460,020
1926	25	8,931	15.75	326	19,549	12,199	10,994	1,539	1,904,766	152,381
1927	16	10,192	6.44	133	12,619	7,155	7,523	986	1,652,186	104,088
1928	12	10,745	6.77	140	25,793	15,089	43,955	6,329	2,209,087	128,127
1929	14	4,618	22.68	469	11,067	5,899	33,382	5,875	1,070,718	67,455
1930	11	6,984	72.31	1,495	1,031	397	18,065	2,348	965,226	48,261
1931	9	3,759	136.27	2,817	341	99	2,865	261	518,086	19,169
1932	7	721	98.27	2,031	16	5	----------	----------	94,795	2,844
1933	2	309	141.76	3,623	535	187	----------	----------	57,927	2,143
1934	7	5,261	989.28	34,575	1,437	928	4,982	399	138,405	5,121
1935	10	9,852	2,011.82	70,414	3,072	2,208	8,435	700	156,013	6,241
Total	-------	504,603	15,587.79	365,875	1,804,597	1,324,706	3,349,475	694,859	82,671,639	5,297,316

5883

I. C. 6964

TABLE 3. - Gold, silver, copper, lead and zinc production from Yellow Pine (Goodsprings) District, Clark County, Nevada, 1902-1935 (Continued)

(Compiled by Charles White Merrill, supervising engineer, San Francisco office, Mineral Production and Economics Division, U. S. Bureau of Mines)

Year	Lode — Zinc — Pounds	Lode — Zinc — Value	Total value	Average recoverable value of ore per ton 1/
1902	----------	----------	$ 44,356	$ 21.35
1903	----------	----------	57,934	27.69
1904	----------	----------	47,963	24.86
1905	685,659	$ 40,454	71,348	29.80
1906	2,885,246	176,000	234,851	24.77
1907	1,878,732	110,845	148,039	33.65
1908	1,115,851	52,445	101,483	21.93
1909	3,013,352	162,721	195,585	22.57
1910	2,707,071	146,182	227,707	38.74
1911	3,548,032	202,238	324,100	37.35
1912	13,254,860	914,585	1,363,354	48.03
1913	14,369,709	804,704	1,239,081	42.64
1914	11,862,149	604,970	864,881	35.25
1915	21,061,182	2,611,587	2,926,300	76.22
1916	28,889,282	3,871,164	4,686,282	65.86
1917	20,535,768	2,094,648	3,294,871	51.27
1918	15,444,731	1,405,471	2,108,195	51.91
1919	5,974,219	436,118	851,976	44.21
1920	9,381,503	759,902	1,196,692	64.30
1921	69,397	3,470	14,081	49.06
1922	891,174	50,797	74,830	40.21
1923	8,755,427	595,369	724,775	49.11
1924	7,909,042	514,088	1,216,077	56.87
1925	6,230,664	473,530	984,827	48.19
1926	2,181,686	163,626	330,071	36.96
1927	3,546,078	226,949	339,311	33.29
1928	1,425,285	86,942	236,627	22.02
1929	1,236,702	81,622	161,320	34.93
1930	2,474,323	118,768	171,269	24.52
1931	1,899,328	72,174	94,520	25.14
1932	254,795	7,644	12,524	17.37
1933	23,346	981	6,934	22.44
1934	28,635	1,231	42,254	8.03
1935	15,380	677	80,240	8.14
Total	193,548,608	16,791,902	24,474,658	48.50

1/ Not to be confused with average assay value.

5883

- 40 -

it is profitable to produce shipping ores. No doubt this camp will become active again if this price is maintained for any appreciable length of time.

The copper ores likewise are replacement deposits in carbonate rocks generally separated from the zinc-lead ores. Production of copper has been restricted to the shipment of higher grade material obtained by selective mining and hand sorting, When the writer visited the district in February 1937 virtually no zinc-lead or copper ores were being mined.

Yellow Pine Mining Co.

The Yellow Pine mining Co., controlled by S. E. Yount of Los Angeles, Calif., owns 16 patented claims in Porphyry Gulch 4 miles west of Goodsprings. This company has been the largest producer of zinc-lead ores in the Goodsprings district. It was originally organized in 1901 by J. F. Kent, and was reorganized in 1906. From 1907 to 1930 it has been a consistent producer. According to the records of the company, production from 1907 to 1925 was 224.245 tons of ore averaging 28 percent zinc, 14 percent lead, and 12 ounces of silver per ton. The net value of this.ore, F;O.B. Jean, Nev., was $27.87 per ton and the total net value was $6,253,583 out of which dividends totaling $3,030,000 were paid. The average cost of mining by the square-set method was $12.11. In 1931 the company suspended operations because of the low price of metals. When the writer visited the property in February 1937, it was under bond and lease to a Las Vegas group and several men were employed in rehabilitating the mine.

Several mills have been built at Goodsprings to treat the ores from the Yellow Pine mine. The first mill was remodeled in 1911 from a mill originally built in 1900 for the leaching of copper ores. This plant separated the lead and zinc products in the coarser sizes by Hartz jigs; the finer materials were classified in a 5-compartment Richards hydraulic classifier, and these products were treated on Overstrom tables. The capacity of the mill was 75 tons per day. This mill was more of a separator than a concentrator, since all the products were marketed except a small amount of waste rock taken off the picking belt. The mill was again remodeled in 1919. and in 1929 it was destroyed by fire. In 1930 a 125-ton flotation mill was erected, but after operating only a short time the mine and mill closed.

Mine development comprises an inclined shaft 900 feet deep and a 300-foot winze from the bottom level. The vertical depth attained in the mine is 465 feet. Total underground workings comprise 52,000 feet. Equipment at the mine includes a 310-cubic-foot-capacity, Chicago, pneumatic, single-stage compressor; a 400-cubic-foot-capacity Ingersoll-Rand Duplex compressor driven by 75-horsepower Fairbanks-Morse Diesel engine; a 50-horsepower Fairbanks-Morse geared hoist; a Sullivan drill sharpener; and all tools necessary for mining. Mill equipment included a Symons cone crusher, Hardinge ball mill, Aiken classifier, 2 Diester double-deck concentrating tables, 6 Fagergren flotation machines, Oliver filter, 2 Dorr agitators, and 2 Dorr thickening tanks. Power for milling is furnished by a 360-horsepower, 4-cylinder, Fairbanks-Morse, Diesel engine.

The ores are all oxidized and occur as irregular replacement deposits in limestone. The deposits have been disturbed considerably by faulting.

Gold and Platinum

The gold deposits in the Goodsprings district are confined mainly to a small area in the vicinity of the Keystone mine. The principal bodies have been found in fractures in limestone or dolomite at or near the contact of granite porphyry. The Keystone mine has been the principal property, with a production of about $600,000, most of which was made in the early days. Small amounts of platinum and palladium associated with copper-gold ore are found in the Boss mine. In recent years most of the mining activity, in this district has been confined to the.gold deposits.

Keystone-Barefoot Mine

The Keystone-Barefoot mine, comprising six patented claims owned by B. H. Kinney, of Albuquerque, New Mex., lies on the west slope of the Spring Mountain Range, 12 miles by road west of Goodsprings via Wilson Summit. The mine was rehabilitated in 1934 after it had lain idle for many years, and in 1935 a 40-ton flotation mill was erected at Sandy 0.7 miles west of the-mine. This mill operated only a short time on ore from the Keystone-Barefoot property. In 1936 the mine was taken over under a 5-year lease by 0. F. Schwartz and five partners, all from Goodsprings. In 1936 the lessees shipped 8 carloads of ore from the Barefoot claim, totaling 286.95 tons and having a gross smelter value of $12,856.90, an average of $44.81 per ton. In addition, several small sack shipments of high-grade ore having a value of about $500 per ton were made. In the early days, principally from 1892 to 1906, the Keystone is reported to have produced $600,000 and the Barefoot $50,000.

Development comprises the Keystone inclined shaft 820 feet deep, the Barefoot inclined shaft 300 feet deep, four adits, and other workings, totaling about 1 mile. Equipment includes a 22-horsepower, Fairbanks-Morse, geared hoist, a 220-foot-capacity, Ingersoll-Rand, portable compressor, a blacksmith shop, and mining tools.

The ore occurs in irregular deposits in limestone at or near a contact of granite porphyry. The values are chiefly in gold in a gangue of iron oxides, calcite, a little quartz, and in places small amounts of oxidized copper minerals.

A representative carload shipment of ore made from the Barefoot claim to the American Smelting & Refining Co. on December 19, 1936, yielded the following data:

Metal quotation:	Au	34.9125 per oz.
Settlement assay:	Au	1.46 oz.
		Percent
	Cu	.52
	Insol	44.6
	Iron	28.3

Metal payment:	1.46 oz. Au at 31.81825		$46.45
Treatment charge:	Base rate	$3.00	
	5 percent excess over $20 gr.	1.32	
		4.32	
	Labor increase	.28	
		4.60	4.60
	Net value per ton		41.85

	Lb.	
Net weight	93,330	
Less 9.2 percent moisture	8,584	
Dry weight	84,716 equals 42.358 tons at $41.85 per ton	
	42.358 tons at $41.85 per ton	$1,772.68
	Less freight advanced $5 per ton	233.25
	Net proceeds	1,539.43

The royalty payments are 20 percent of the net smelter returns.
The hauling cost to Jean 20 miles from the mine is $3 per ton on contract.

Boss Mine

The Boss mine, comprising a group of five patented claims owned by S. E. Yount, of Los Angeles, Calif., and associates, lies on the west side of the Spring Mountain Range, 14 miles by road west from Goodsprings. Originally located as a copper prospect in 1886. it has been worked intermittently as a source of copper, gold, and platinum. In 1900, a company erected a leaching plant at Goodsprings for the treatment of copper ores from the Boss and Columbia mines, but it was unsuccessful. In 1914, platinum and palladium were discovered in the Boss mine by H. C. Riddall. Considerable publicity was given to this discovery, and the high assays of selected platinum ore gave the impression that it was an unusually high-grade mine, which was not the case. In 1914 the Boss Gold mining Co. was organized, and in the same year the property was sold to a syndicate headed by W. C. Price, of Los Angeles, which organized the Platinum Gold Mining Co. In 1915 the property reverted to the Boss Gold Mining Co. In 1936, after lying idle for many years, the mine was leased to Roy Myers, and in 1937 the lease was transferred to E. E. Kinney.

According to Hewett[13], production of the mine up to 1920 was 3,500 tons of ore having a gross value of $210,000. The most productive period was from 1914 to 1920, when 3,051 tons of crude ore were mined containing 1,771.81 ounces of gold, 7,014 ounces of silver, 396 ounces of platinum, 594 ounces of palladium, 568,099 pounds of copper, and 753 pounds of lead.

From March to November 1936, Myers shipped 261.87 tons of ore having a gross value of $5,086.81, an average of $19.42 per ton. This ore was obtained by screening the dumps. From December 1, 1936, to March 1937 Kinney treated 1,200 tons of dump material in the Keystone mill.

Mine is developed by five adits and other workings, totaling about 1 mile. There is no mining equipment on the property. An aerial, 2-bucket, gravity-type tram 1,000 feet long connects the upper workings with a loading bin.

The ore occurs in dolomitic limestone along a fault zone striking about northeast and southwest. Most of the ore was derived from a stope about 150 feet long, 25 feet wide, and 25 feet high above the upper tunnel level. This ore body was an irregular mass of quartz and iron oxides containing copper minerals, gold, silver, and a small amount of platinum and palladium.

A carload of screened dump ore shipped by Roy Myers on November 28, 1936, to the International Smelting and Refining Co. yielded the following data:

Metal quotations:	Au	$34.9125 per oz	
	Ag	.77 per oz	
	Cu	10.15 less 2.525 cents per pound	
		Oz.	
Settlement assay:	Au	.67	
	Ag	1.8	
		Percent	
	Cu	2.02	
	Insol	30.0	
	Fe	6.3	
	CaO	15.4	
Metal value	Au at $31.81825		$21.32
	Ag at $.77		1.32
	Cu less .5 at 1.525		2.32
		Total	24.96

Geologist D. F. Hewitt of the U.S. Geological Survey captured this view of Goodsprings only months after the flood of 1916. Though Goodsprings was in the center of numerous mining operations, its business district (shown leading off the picture in the foreground right) never developed to a large degree.

Treatment charge:	Base charge	$3.50		
	10 percent excess over $20 gr.	.50		
		3.50		
	Labor increase	.34		
		3.84	3.84	
	Net value per ton		21.12	
	Lb.			
Wet weight	75,100			
Moisture, 4.0 percent	3.004			
Dry weight	72,096 or 36.045 tons			
	36.045 tons at $21.12			$761.33
Deductions:	Freight $4.50 per ton			168.98
	Net proceeds			592.35

The hauling cost to Jean, 22 miles from the mine, is $3 per ton on contract.

Royalty payment is 10 percent of the net smelter or mill return.

The flow sheet of the Kinney custom mill is shown in figure 6. It is at Sandy 3 miles southwest of the Boss mine. At the time of the writer's visit the mill was operating entirely on dump material from the Boss mine. Capacity of the mill is 40 tons per day. Power is furnished by a 55-horse-power Ingersoll-Rand Diesel engine and a 20-horsepower Fairbanks-Morse Diesel engine, the latter connected to a 25-kv-a generator. Water for milling is obtained from a well near the millsite.

Chiquita Mine

The Chiquita property, comprising 18 claims (1 patented) owned by the Chiquita Mining Co., Ltd., James Maxfield, general manager, lies 12 miles by road west of Goodsprings via Wilson Summit. In the early days considerable work was done on the Chiquita and Hillside claims, but there is no record of any production. In April 1931 0. F. Schwartz and six associates, all from Goodsprings, obtained a lease and bond for $10,000 on the Chiquita and Hillside claims from the owner, A. J. Robbins of Goodsprings. The lessees discovered a body of milling ore and, after working it 1 1/2 years, sold their bond and lease to James Maxfield of Los Angeles for $80,000. The purchase price included a group of claims the lessees had located after the ore was discovered. Development work was continued by the Chiquita Gold Mining Co. Ltd., and in July 1936 a 100-ton all-slime cyanide plant was completed. When the writer visited the property 70 men were employed in the production of about 80 tons of ore per day averaging about $12 per ton.

The mine is developed by an inclined shaft 750 feet deep and other workings totaling about 2 1/2 miles. Property is completely equipped, and recently a 350-horsepower Fairbanks-Morse Diesel generating set has been installed. After for milling is obtained from a well in Mesquite Valley and is pumped to the mill through a 4-inch-diameter pipeline 6 miles long. Total lift is 2,000 feet.

Values are chiefly gold, which occurs in irregular bodies in limestone near granite porphyry.

Red Cloud Group

The Red Cloud group of two unpatented claims owned by John D. Loop, of Goodsprings, lies about 4 1/2 miles northwest of Goodsprings. This property was originally located in 1902, and in 1906 a small cyanide leaching plant was erected at the mine to treat the ore. According to Loop, about 2,000 tons of ore were cyanided and $10,000 in gold recovered. In 1934 the property was leased to P. S. McClanahan, who shipped 36.66 tons having a gross value of $1,726.32, an average of $47.09 per ton. In 1936 Loop shipped 26,193 tons having a gross value of $679.18, an average of $25.93 per ton. The foregoing constitute the total production.

The main shaft on the property is vertical to a depth of 116 feet, from which point it is inclined 68 degrees to a depth of 300 feet. Underground workings total 2,000 feet. The only equipment on the property is a gasoline hoist and a blacksmith shop.

The ore occurs in irregular bodies at or near the contact of a granite porphyry dike that cuts the dolomitic limestones. Values are chiefly in gold in a gangue of silicified material stained with iron oxides. An interesting feature of the property is that cinnabar was found on the upper levels associated with gold. The cinnabar, however, has not been commercially important. The smelter returns on the shipment of ore made by Loop on October 14, 1936, to the American Smelting and Refining Co. yielded the following data:

Metal quotation:	Au	34.9125			
		Oz			
Settlement assay:	Au	.815			
	Ag	.1			
		Percent			
	Insol	86.4			
	Fe	3.6			
	CaO	2.6			
Metal value:	0.815 oz. Au at 31.81825			$25.93	
Treatment charge:	Base		$3.25		
	10 percent excess over $20 gr.		.59		
			3.84		
	Labor increase		.28		
			4.12	4.12	
	Value per ton			21.81	
	Lb.				
Wet weight	53,620				
Less 2.3 percent moisture	1,234				
Dry weight	52,386	26.193 tons at $21.81			$571.27
Deductions:	Freight, $4.50 per ton				120.65
	Net proceeds				450.62

The hauling cost to Jean, 12 1/2 miles from the mine, is $2 per ton on contract.

Other Mines

The Clementina group of four unpatented claims, owned by H. A. Hogle of Salt Lake City, Utah, lies 1/2 mile southeast of the Keystone mine. At the time of the writer's visit this ground was being prospected by two men. Only a few tons of ore have been shipped from the property in former years.

Development work comprises a shaft 60 feet deep and an adit 400 feet long. Equipment includes a Fairbanks-Morse compressor (8 by 8 inches) driven by a 16-horsepower International tractor engine. Gold ore occurs in pockets in a shear zone in dolomite.

The Yellow Horse group of two patented claims and the Honduras group of four patented claims are controlled by the Minerals Exploration Co. At the time of the writer's visit this company was exploring the Yellow Horse property with a crew of three men. There is no record of any production. Development work comprises an inclined shaft 60 feet deep with about 200 feet of lateral workings. Equipment includes a 6-horsepower Fairbanks-Morse geared hoist and a one-drill Fairbanks-Morse air compressor belt-connected to a 26-horsepower Western gasoline engine. The formation is limestone intruded by granite porphyry.

Molybdenum

Molybdenum in the form of wulfenite ($PbMoO_4$) is found in the Shenandoah, Milford, Hoosier, Mobile, Hermosa, and other mines in the district. The wulfenite occurs in small irregular deposits in limestone associated with zinc-lead ores. The only effort made to produce molybdenum concentrates was in 1935-36 when the California Molybdenum Corporation acquired the Shenandoah mine under bond and lease and erected a mill near Sandy, 3 miles from the mine. Judging from the tailings pile at the mill, about 5,000 tons of wulfenite ore were treated, but no information is available on the amount or grade of concentrates produced. The Shenandoah mine and mill reverted to the original owners in November 1936.

Shenandoah Mine

The Shenandoah mine, owned by P. H. Springer of Las Vegas, Nev., and J. A. Fredrickson of Goodsprings, comprises a group of 13 claims (1 patented) on the west side of the Spring Mountains, 16 miles by road west from Goodsprings. This property l.as been intermittently worked for zinc-lead ores, principally by lessees, for many years. According to Hewett[14], production from 1908 to 1926 was 1,143 tons of shipping ore containing 2,120 ounces of silver, 208,580 pounds of lead, and 538,557 pounds of zinc.

The mine is developed by several adits, which, with lateral workings, total about 2,500 feet. The mine is near the top of a steep mountain, and an aerial gravity tramway 1,650 feet long connects the mine with the loading bins. Mine equipment has been removed from the property.

Wulfenite is present in lenses in brecciated zones in limestone. According to Fredrickson, the ore mined by the California Molybdenum Corporation was found in a block of ground 130 feet long, 160 feet high, and from 10 to 30 feet wide. The zinc-lead ore minerals are cerussite, galena, smithsonite, and hydrozincite.

The Shenandoah mill has a capacity of 50 tons of wulfenite ore per day. Concentration was effected by tabling followed by flotation. Mill equipment includes an Eimco ball mill (4 by 5 feet), Dorr simplex classifier, 2 Deister tables, trommel (5 by 2 1/2 feet, equipped with 8-mesh screen), 12 M. S. flotation cells, an Eimco filter, and other equipment. Primary crushing was done at the mine. The power plant at the mill consists of a 180-horsepower, 4-cylinder, Diesel engine connected to a 170-kv-a alternator and a 75- horsepower, Fairbanks-Morse, 2-cylinder, Diesel engine connected to a 50-kv alternator. A power line connects the mine with the power plant.

Water for milling is available at the millsite from a well 186 feet deep, lined with 6-inch casing, and equipped with a 52-gallon-per-minute-capacity, Pomona, deep-well, turbine pump connected to a 5-horsepower motor. The corrugated iron stand tank near the well has a capacity of 16,000 gallons.

Vanadium

Vanadates of lead, zinc, and copper have been found on the Spelter, Hoodoo, Whale, Valley Forge, Velvet, Akron, and a number of other properties in the Goodsprings district. All of the vanadium deposits in the district were found while prospecting was being done for zinc-lead ores.

The production of vanadium has been about 50 tons of hand-sorted shipping ore containing about 10 percent V_2O_5, and 200 tons of milling ore averaging about 1.6 percent V_2O_5. The largest shipment was made in 1918 by Charles Kennedy from the Lookout mine on the east side of the Spring Mountain Range. This shipment contained 36 tons and averaged 9.26 percent V_2O_5. In 1923 an unsuccessful attempt was made to concentrate vanadium ore on the Spelter property by gravity methods. In 1936, 182 tons of vanadium ore mined from the Spelter and Hoodoo properties ore treated at the Kinney custom mill by flotation. From the 182 tons of ore averaging 1.65 percent V_2O_5, 6.6 tons of concentrates were produced containing 16.7 percent V_2O_5, 43.3 percent lead, 3 percent copper, and some zinc. Since the recovery of vanadium on this test run was only 36.8 percent, the mill would not accept further shipments of vanadium ores.

All the vanadium deposits are in brecciated zones in limestone or dolomite. The principal vanadium minerals are vanadinite ($3Pb_3V2O_8.PbC_{12}$), descloizite ($4RO,V_2O_5.H_2O$;R=Pb,Zn, usually in the approximate ratio 1 to 1), and cuprodescloizite ($(Pb,ZnCu)(V0_4)2(Pb,Zn,Cu)(OH)_2$), associated with limonite, galena, and oxidized lead and zinc minerals.

In general, the vanadium ores are too low-grade to be shipped crude. with the present stage of development, no single property in this area is capable of producing sufficient vanadium ore to keep a 25-ton daily capacity concentrator operating steadily, However, the writer is of the opinion that if the properties were worked by lessees they could, together, supply sufficient-ore averaging better than 1 percent V_2O_5 to keep a 25-ton mill busy. The mill would have to be operated on a custom basis. The various economic factors involved in an enterprise of this nature would have to be investigated further.

Laboratory milling tests made on vanadium ores from the Goodsprings district by the U. S. Bureau of Mines Experiment Station at Rolla, Mo., indicate that the ores are amenable to concentration by the flotation process. In continuous, closed-circuit milling a recovery of 80 to 85 percent of the vanadium minerals appears feasible.

Because of the complex nature of the vanadium minerals the concentrates require a special treatment process to reduce them to finished products. The U. S. Bureau of Mines Experiment Station at Reno, Nev., has made some tests on the reduction of the vanadium concentrates along the lines of caustic fusion, direct smelting, and leaching with sulphuric acid. The interested reader is referred to the report on the results of these tests made by Doerner[15].

Smelter Group

The Smelter Group of four unpatented claims, incorporated under the name of Master Alloy mining and Milling Co., Rex C. Ewing, Goodsprings, Nev., principal owner, lies 9 miles southwest of Goodsprings on the west slope of the Spring Mountain Range. These claims were located originally for lead and zinc, but there is no record that these minerals were pro- duced. In 1923 an attempt was made to concentrate the vanadium ore in a small mill equipped with crusher, rolls, screens, and four Wilfley tables. About 3 1/2 tons of vanadium concentrates averaging 11 percent V_2O_5 were produced and sold to Elmer H. Hester of Kingman, Ariz. In 1936, 72 tons of vanadium ore averaging 1.84 percent V_2O_5 were treated in the Keystone Custom mill at Sandy.

Property is developed by three tunnels which, with lateral workings, total about 1,000 feet. There is no equipment on the property.

The ore occurs in a brecciated zone in limestone striking east and west and having a maximum width of 25 feet. The vanadium occurs as vanadanite and descloizite associated with galena. Lenses of ore up to 6 feet in width are said to contain from 1 to 2 percent V_2O_5. Two cut samples taken by the writer over widths of 20 and 25 feet-analyzed .25 and .45 percent V_2O_5, respectively.

Hoodoo Group

The Hoodoo group of four unpatented claims, incorporated under the name of the Kansas Nevada Mining Co., controlled by Frank Williams of Goodsprings, adjoins the Spelter group on the southeast. This property was located originally for zinc-lead, and it has been worked intermittently for a number of years on a small scale. According to Hewett[16], the production of crude zinc..lead ore from 1911 to 1927 was 335 tons containing 523 ounces of silver, 33,831 pounds of lead, and 189,954 pounds of zinc. According to Williams, 16 tons of vanadium ore, averaging 4 percent V_2O_5 were sold to Elmer H. Hester, Kingman, Ariz., in 1928. In 1936 110 tons containing 1.52 percent V_2O_5 were treated in the Keystone custom mill at Sandy.

Development work comprises three adits and other workings totaling about 2,000 feet. There is no equipment on the property.

Vanadates occur in lenses in a shear zone in limestone. Three cut samples taken by the writer over widths from 5 to 9 feet contained .0.67, 0.68, and 1.18 percent V_2O_5.

Whale Group

The Whale group of six unpatented claims owned by Frank Miller of Goodsprings lies on the west slope of the Spring Mountain Range, 8 miles by road west of Goodsprings. This property has been worked intermittently for zinc-lead ore. According to Hewett[17], from 1912 to 1917, 537 tons of crude ore were shipped, containing 265,868 pounds of zinc, 21,238 pounds of lead, and 98 ounces of silver. The first discovery of vanadium ore was made on this property by Frank Miller in 1912 while prospecting for zinc-lead. No vanadium ore has been produced.

A deposit of zinc-lead has been prospected by an adit and subsidiary workings totaling about 500 feet. The vanadium deposits have been prospected by two shallow shafts, several short adits, and numerous open-cuts totaling about 250 feet of workings. The latter work on vanadium deposits was done while prospecting for lead-zinc ore.

The vanadium occur chiefly as cuprodescloizite in shear zones in limestone. Small amounts of galena, calamine, and wulfenite are present. Vanadium showings have been found in irregular deposits in at least 20 places scattered over an area 60 feet wide and 1,000 feet long. There is no continuity to the deposits. A general sample of dump material taken by the writer from 10 open-cuts analyzed 1.12 percent V_2O_5 and another sample of dump material near a shallow shaft analyzed 0.73 percent V_2O_5. By selective mining the grade of the ore probably could be maintained above 1 percent vanadium pentoxide.

Valley Forge Group

The Valley Forge (formerly known as the Hickory Dick) group of four unpatented claims, owned by Frank Williams o. Goodsprings, lies 2 miles west of Columbia Summit and 6 miles southwest of Goodsprings. This property was located originally for copper, and during the World War four carloads of copper ore was mined and shipped. Vanadium ore has not been produced.

Development work on the copper showings comprises 250 feet of adits and other workings. The vanadium deposits have been prospected by five open- cuts totaling not more than 50 feet. There is no equipment on the property.

Vanadates (chiefly cuprodescloizite) occur in brecciated zones in lime- stone over widths of 12 to 15 feet. Several samples were taken by the writer; one sample of dump material analyzed 0.72 percent V2O5 and a cut sample taken over a width of 3 feet on the hanging-wall side of a deposit

15 feet wide analyzed 1.26 percent V_2O_5. With a small amount of work done, the showings are encouraging.

Velvet Group

The Velvet group of three unpatented claims, owned by Charles Kennedy of Goodsprings, lies on the north side of Table Mountain, 8.7 miles by road southwest of Goodsprings. Nothing has been produced from the property.

Development comprises an adit 50 feet long and other scattered workings totaling about 100 feet. Vanadates, principally vanadinite and descloizite, occur on a shear zone in limestone. A channel sample taken by the writer across a width of 3 feet in the face of the adit analyzed 0.79 percent V_2O_5.

Akron Group

The Akron Group of three patented claims, owned by Charles Beck of Goodsprings, adjoins the Valley Forge group on the west. The Bill Nye claim, formerly a separate property, constitutes part of the Akron group. These claims originally were located for zinc-lead. According to Hewett[18], production from the Bill Nye claim from 1914 to 1919 was 1,488 tons of crude ore containing 6,860 pounds of copper, 62,564 pounds of lead, 921,931 pounds of zinc, and a little silver and gold. Production from the Akron claim has been four carloads of zinc-lead-silver ore and 7 tons of vanadium ore containing about 7 percent V_2O_5.

Development comprises several shafts, the deepest of which is 220 feet inclined 76, and other workings totaling about 1,000 feet. Virtually all the development work has been done in the search for zinc-lead ore. There is no equipment on the property.

Vanadates, chiefly cuprodescloizite, occur in brecciated dolomitic limestone on the Akron claim.

Cobalt

Cobalt, in the form of stainierite $(Co,Fe,Al)2O_3.H_2O$, occurs in a number of mines in the Goodsprings district. The only production was made in 1921-22, when a number of small lots aggregating 21 tons, containing 5.13 to 29.18 percent and averaging 8.73 percent cobalt, were shipped from the Blue Jay, Columbia, Copper Chief, and Highline properties. Although the presence of cobalt is rather widespread in the area, the deposits so far described have been small. In recent years the only prospecting done for cobalt was on the Swansea property.

Swansea Group

The Swansea group of three unpatented claims, owned by Frank Williams of Goodsprings, lies about 1 mile west of Columbia Summit. This property was located originally for copper, and 4 years ago it was relocated for cobalt. None has been produced.

Development consists of several short adits totaling not more than 100 feet of workings. Where is no equipment on the property. The stainierite occurs as irregular replacement deposits in limestone. It is associated with considerable calcite. About 4 tons of ore on the dump was reported to contain 3 1/2 percent cobalt.

JEAN DISTRICT
(Silica Sand, Radium, Limestone)

Silica sand, radium (carnotite), and limestone occur in the vicinity of Jean, a station on the Union Pacific R.R. 30 miles by road a little west of south of Las Vegas, Nev.

The only production from the district has been small amounts of silica sand and limestone. The carnotite occurrences have been of more mineralogical than commercial interest.

Nevada Pacific Minerals Co.

A large deposit of silica sand occurs in T. 24 S, R. 60 E., M.D.M., 2.7 miles northwesterly from Sutor Siding. Sutor Siding is on the main line of the Union Pacific R.R. about 4 1/2 miles north of Jean, Nev.

In August 1936 the Nevada Pacific Minerals Co. of Los Angeles, Calif., E. S. McCurdy, president, started to work the deposit, and up to January 1937 had shipped 25 cars of silica sand for molding purposes. Due to lack of water in the vicinity of the deposit no silica sand has been produced for the better grades of glassware. An average of three men is employed.

The sand occurs in a sandstone formation exposed in a well-rounded hill. The sandstone bed mined consists largely of white quartz grains stained with iron oxide on the bedding planes and in the seams. At the surface the sandstone is cemented with lime.

Mining is done by open-cut method. The sandstone is fairly well consolidated, and jackhammers with detachable Carr bits and 7/8-inch hollow hexagonal steel are used for drilling. Blasting is done with 40-percent gelatin-dynamite No. 6 detonators and tape fuse. Compressed air is furnished by a portable Ingersoll-Rand compressor. Water for mining and domestic purposes is hauled from Las Vegas in railroad cars to Sutor Siding and then trucked to the camp.

The broken sandstone is shoveled by hand into wheelbarrows equipped with rubber tires and transported a short distance to a bin. From the bin it is fed to a Blake type crusher (8 by 15 inches) set to crush to 1/4-inch size. The crusher discharges into a boot of a belt and bucket elevator that discharges into a trommel equipped with 1/4-inch and 16-mesh screens. The plus 1/4-inch and minus 16-mesh products, which constitute about 10 percent of the material mined, is rejected as waste. The minus 1/4-inch and plus 16-mesh products drop into a bin, from whence they are loaded into trucks and hauled to the railroad. The sandstone is crushed and sized dry. Power for the plant is furnished by a 30-horsepower Waukesha semi-Diesel engine.

Radium Deposits

Carnotite was discovered several miles west of Sutor Siding in 1931 by J. Isachsen, formerly of Jean, Nev. Following Isachsen's discovery, other deposits were located in the vicinity.

The carnotite, associated with manganese oxide, occurs as a coating in fractures in sandstone and limestone. Although prospecting has been carried on by a number of shallow shafts, trenches, and open-cuts, the amount of carnotite found was too small to be of commercial importance.

Limestone Deposit

A limestone quarry about 1 mile southeast of Jean, Nev., on Sheep Mountain, was opened by the Blue Diamond Gypsum Co. about 10 years ago. After 850 tons of lime rock had been shipped, the quarry was abandoned.

LAS VEGAS DISTRICT
(Manganese)

The Las Vegas district is 16 miles southeast of Las Vegas, Nev. Manganese was discovered here in September 1917 by R. N. Edwards, B. R. Jefferson, and J. F. Marrs of Las Vegas, who located the Three Kids group of claims. Shortly after, a local group formed the Manganese Association, and production was begun in November 1917. In 1918 the Three Kids mine and the adjoining Las Vegas group were purchased by Thomas Thorkildson of Los Angeles. Calif., but the name of the company was retained. The Manganese Association worked the property until March 1919. The property remained idle until 1936, when it was acquired by Los Angeles interests, who formed the Boulder Dam Manganese Co. Production during the war was about 16,000 tons of shipping ore that averaged 40 percent manganese.

Interest in these manganese deposits was revived in 1936 because of the research work carried on by the U. S. Bureau of mines on the electro-deposition of metallic manganese from an electrolyte obtained by leaching manganese ore[19].

Boulder Dam Manganese Co.

The Boulder Dam Manganese Co., H. R. Golenor of Los Angeles, president, and G. R. Boggs, of Los Angeles, vice president and consulting engineer, controls the principal properties in the Las Vegas manganese district, com- prising 11 claims, two of which are patented.

Development work comprises a number of shallow shafts and inclines, none over 100 feet deep, and several short tunnels. There is no equipment on the property. In February 1937 the property was idle.

During the World War mining was done by either open-cuts or the room- and-pillar system. Where the overburden was comparatively thin it was broken up and removed by horse-drawn scrapers and the ore was mined by open-pit methods; where the overburden became too thick to be handled the room-and-pillar system of mining was employed. According to Hale[20] the first 31 carloads of ore shipped from the Three Kids property averaged as follows:

	Percent
Mn	41
SiO_2	11
P	0.03
Fe	1
Moisture	11.62

In 1936 the Boulder Dam Manganese Co. explored its holdings by diamond drilling. As a result of this work, ore was blocked out as follows:

	Tons ore (2,000 lb.)	Mn content, percent	Ave. thickness, feet
Block A	83,000	35.33	34
Block B	228,000	30.5	37
Block C	65,000	18.5	20
Three Kids	100,000	31.0	20
Total	476,000	29.81	

The metal content of the ore blocked out is 141,889 tons of metallic manganese.

The geology of the Las Vegas manganese deposit has been described by Pardee and Jones[21] and Hewett and Webber[22].

According to Hewett and Webber, the Las Vegas manganese deposits are of sedimentary origin. The formation in the vicinity consists of numerous volcanic flows, breccias, and tuffs. The overlying stratified materials that contain the beds of manganese oxides are largely reddish-brown tuffs, with conglomerate of volcanic material and gypsum beds. Drill records show the average thickness of the overburden to be about 26 feet.

The manganese oxides, chiefly wad and psilomelane, occur as soft porous material weighing about 80 pounds per cubic foot. The thickness of the bed varies from a few feet to a maximum of 60 feet. The bed has been disturbed by faulting. The ore is homogeneous, and a high-grade product cannot be obtained by sorting or selective mining. Lenses of sandy material and seams of fibrous gypsum ranging in thickness from a few inches to 2 feet occur in the manganese bed. This material is easily separated from the ore in mining.

Ebony Queen Claim

Manganese occurs 2 miles northwest of Boulder City..Nev., near Hemmingway Pass on the Ebony Queen unpatented claim relocated by J. R. Sailor and G. H. Miller of Las Vegas, Nev., in January 1937. This deposit was explored about 1925 by several open-cuts totaling about 50 feet of work. Nothing has ever been produced.

The deposit striking northwest and southeast and dipping 75° out-crops for several hundred feet along the side of a small hill. Material consists of a mixture of iron and manganese oxide ranging from 4 to 10 feet in width. No information on the manganese content is available.

MOAPA DISTRICT
(Gypsum, Volcanic Ash)

The Moapa district is in the vicinity of Moapa, Nev., a station on the Union Pacific R.R. that connects Los Angeles, Calif., and Salt Lake City, Utah. Gypsum and volcanic ash occur in this area but the only production has been gypsum.

Gypsum Deposits

From 1919 to 1923 the White Star Plaster Co. exploited a gypsum deposit 3 1/2 miles east of Moapa. The White Star mill originally contained two lO-foot, oil-fired, calcining kettles, but the plant was enlarged to be a 3-kettle mill in 1922. During 1922 shipments of plaster to Los Angeles, Calif., averaged about 3,000 tons per month[23].

In 1923 the company ceased operations, and no work has been done on the deposits since.

Two beds of gypsum occur in limestone, but only the lower bed was exploited. This bed is from 30 to 40 feet thick, dips about 30°, and is exposed for several thousand feet.

In the Muddy River Valley, about 5 miles south of Moapa, lies the Anderson gypsum deposit. Strata of gypsum interbedded with shale outcrop for a distance of 3 miles. within a gypsiferous zone several hundred feet thick are strata of rock gypsum apparently containing no anhydrite. The principal gypsum bed is from 10 to 60 feet thick, and it could be mined at low cost.

The nearest railroad point is Jackson Siding, 4 miles east of the deposit, on the St. Thomas branch of the Union Pacific R.R. About 20 years ago the Rex Plaster Co. shipped a small amount of gypsum from this deposit to a plant in Los Angeles.

Volcanic Ash Deposit

A deposit of volcanic ash owned by the Dotson Bros., Mart and Henry, of Logandale, Nev., occurs 13.6 miles northwest of Moapa. The deposit is covered by four unpatented lode claims.

The volcanic ash occurs as a horizontal bed about 30 inches thick, exposed on a side of a bluff for a distance of about 1 mile. The material is uncemented, fine-grained, and grey in color. No effort has been made to work the deposit.

Tin Deposit

In 1936 tin was reported to have been found about 13 miles southeast of Moapa, Nev. The Cosmo Mining Co. acquired a large group of claims and some stock was sold. Pieces of sandstone stained with manganese, said to contain tin, were taken by the writer. A composite sample of the material was analyzed by the Mackay School of Mines laboratory at Reno, Nev., but no tin was found.

MUDDY MOUNTAINS DISTRICT
(Borates, Gypsum, Silica Sand Bentonite, Sodium Sulphate)

Deposits of borates, gypsum, silica sand, bentonite, and sodium sulphate occur in the southern part of the Muddy Mountains in eastern Clark County north of Lake Mead. The only production of any consequence has been from the borate deposits. The non-metallic deposits that occur in the northern part of the Muddy Mountains are described in this paper under the Moapa and St. Thomas districts.

The borate deposits, which occur chiefly as colemanite ($Ca_2B60_{11}.13H_20$), are of interest in that they constitute the only commercial occurrences of this mineral in the United States outside of California. These deposits were discovered in 1920 and 1921 and were exploited from 1922 to 1928. In the latter year operations ceased, due to the discovery of a new boron mineral, kernite ($Na_2B40_7.4H_20$), near Kramer, San Bernardino County, Calif. Because of the different types of borate deposits that have been worked successively in the United States, the borax industry has been very unstable, and at every stage of development different processes of exploitation were devised.

In the United States the industry began with the discovery of borax ($Na_2B4O_7.10H_2O$) by John A. Veatch in the waters of Tuscan Springs, Tahama County, Calif. Doctor Veatch extended his search, and other deposits were found in the same year on the east side of Clear Lake, Calif. The borax from hot springs and Clear Lake constituted the chief source of supply in the United States for many years. In the early seventies, when the lake deposits of California were virtually exhausted; borax and ulexite ($NaCaB_5O_9.H_2O$) were discovered in the desert playas in Nevada and California. These playa deposits were worked up to 1887, when the industry was revolutionized by the discovery.of colemanite beds near Death Valley, San Bernardino County, and later in Inyo and Los Angeles Counties, Calif. The latest development in the domestic borax industry was the discovery of extensive deposits of kernite, which supplanted colemanite in much the same manner as colemanite supplanted the playa deposits of borax and ulexite. Kernite has a smaller content of water of crystallization than the borax of commerce, so that when it is recrystallized to borax the original weight is increased.

Borate Deposits

Borates occur in two localities 13 miles apart in the southern end of the Muddy Mountains. One deposit is in the White Basin, 20 miles by automobile road southeast of Crystal, Nev., and the other is near Callville wash, 26 miles by road east of Lovell Siding on the Union Pacific R.R.

The White Basin colemanite deposit was discovered by John F. Perkins of Overton, Nev., in 1920. Shortly after, the Perkins group of 12 claims was sold to the Pacific Coast Borax Co. Subsequently other locations were made in the vicinity, which were acquired by the American Borax Co. This company operated from 1922 to 1924, when operations ceased because of litigation with the Pacific Coast Borax Co. Since 1924 there has been no borax activity in this area.

Development work done by the American Borax Co. includes an inclined shaft and several hundred feet of lateral workings. The calcining plant erected by this company near the mine has been dismantled.

According to Noble[24], colemanite occurs in bunchy lenticular layers interbedded with white shale containing gypsum, beds of volcanic ash, or thin layers of limestone. Some of the limestone shows a peculiar concretionary structure called "goose eggs," and these concretions are believed by the prospectors to indicate the presence of boron minerals. The colemanite beds vary in thickness from a few inches to 3 feet.

Stimulated by the discovery of borates in the white Basin, F. M. Lovell and George D. Hartman of St. Thomas, Nev., who were prospecting under a grubstake agreement, discovered the Callville wash colemanite deposit on January 23, 1921. As soon as it became known that the deposit was valuable, a number of intending purchasers tried to purchase the property, but they were outplayed by F. M. (Borax) Smith, who, in the early days. was the dominating personality in the borax industry of the world. Smith, after examining the property, made an offer and arranged an initial deposit of $50,000 to bind the bargain. The contract was drawn up in the field on the leaf of a notebook, and the purchase price is said to have been $250,000. In July 1922 the West End Chemical Co., of which Smith was the leading figure, established a camp and began production. In 1928 operations ceased because the colemanite could not compete with the kernite deposits mined by the Pacific Coast Borax Co. near Kramer, Calif.

According to J. W. Wilson of Las Vegas, Nev., formerly superintendent of the property (which is known as the Anniversary mine), the production of crude colemanite averaging about 20 percent B2O3 from 1922 to 1928 was about 200,000 tons. During the period of greatest activity 175 tons of crude ore were mined per day, and from this amount of ore about 65 tons of calcines were obtained averaging about 40 percent B2O3. The calcines were hauled 24 miles to the railroad and shipped to refineries on the Pacific coast.

The West End Chemical Co. owns a group of 21 lode claims, 12 of which are patented. The lode claims are also covered by seven placer locations.

Entry to the Anniversary mine is made by two adits. Underground workings comprise approximately 7,000 feet. An electrically operated aerial tramway 12,000 feet in length was employed for transporting the ore from the portal of the main adit to the calcining plant, 250 feet lower than the adit elevation. Calcining equipment comprises two rotary, oil-fired, double- tube kilns, each 60 feet long and 6 feet outside diameter. The inner chamber of each kiln is 3 feet in diameter. Indirect firing was employed in the kilns. The annular space between the two tubes constituted the firing chamber. Power for mining and milling was furnished by a 4-cylinder, Fairbanks-Morse, Y-type, Diesel engine connected to a generator. The mine workings, camp, and equipment are still intact and in fairly good condition. It is reported that nearly $2,000,000 was invested in this property.

The geology of the Anniversary mine has been described by Noble25 and by Callaghan and W. W. Rubey[26]. The colemanite bed dips from 45° to 60°, averaging 50°, and is traceable on the surface for a distance of about 3,000 feet. It is interbedded with shale and limestone in a synclinal basin whose axis trends east and west. The width of the bed ranges from 6 to more than 22 feet, the average minable width being about 8 feet. According to Callaghan and Rubey, the ore reserves are at least 400,000 tons.

The shrinkage-stope method of mining was employed. Due to the fact that colemanite breaks into tabular pieces, its angle or repose is greater than that usually encountered in metalliferous deposits. Where the bed was steep enough, the stopes were kept filled with broken colemanite, and in the flatter portions as much ore as possible was retained in the stopes and small pillars or ore were left for support at irregular intervals. The only timber used in mining was for manways and chutes, the latter spaced at intervals of 20 feet. where the colemanite bed was too flat to permit the ore to run readily, the final cleaning up of the stones was effected by scrapers. About 90 percent of the ore was recovered in mining. The ore was trammed by hand in 1-ton cars to the tramway loading terminal at the portal of the main adit.

At the calcining plant the ore was crushed to 1-inch size and fed to the two rotary kilns. The calcined material was discharged on an electric vibrating screen, which separated the colemanite from most of the associated impurities. The separation of colemanite from the impurities is possible by screening because of the fact that the former disintegrated to a fine powder while the latter do not. The screened product, averaging about 40 percent B_2O_3, was hauled in trucks to the railroad for shipment.

Gypsum Deposits

Gypsum occurs in the Muddy Mountains about 2 1/2 miles southwest of Comstock, Nev. The deposit is 3 1/2 miles from the Union Pacific R.R. C. C. McDonald, of Overton, Nev., and associates own 320 acres, and the Pacific Portland Cement Co. owns 640 acres, the latter purchased from McDonald in 1922. Prior to purchasing the property, the Pacific Portland Cement Co. explored the deposit by diamond drilling. The gypsum bed is 150 feet thick and dips 45°. At a depth of about 100 feet the gypsum changes to anhydrite. Analyses indicate that the material is nearly pure.

Deposits of gypsum and gypsite occur several miles from Lake Mead on the east side of Boulder wash, which drains into Lake Mead at the east end of Boulder Canyon. The gypsum and gypsite outcrop over an area of several square miles, but no attempt has ever been made to exploit the deposits because of their inaccessibility. When the water in Lake Mead rises to the crest of Boulder Dam the gypsum can be transported readily by barge to Boulder City, 12 miles distant.

Silica-Sand Deposits

An extensive deposit of white silica sand occurs on the west side of the Muddy Mountains, 8 miles southwest of Crystal, Nev. The deposit is covered by the Lucy Etta group of five unpatented placer claims, totaling 800 acres owned by C. D. Wyatt of Boulder City,, Nev. The former loading station of the American Borax Co. on the Union Pacific R.R. at Crystal is 8 miles from the deposit. The freight rate to Los Angeles, Calif. is $2.30 per ton and to San Francisco $3.60 per ton. No water is available for a washing plant in the vicinity of the deposit. The only production has been four carloads of sand shipped, without treatment, to the Technical Glass Co. of Los Angeles in 1934, that analyzed as follows:

	Percent
Calcium	0.042
Iron	.022
Aluminum	.22
Magnesium	.045
Silica	99.04

Both white and pink varieties of sand occur on the Wyatt claims in a sandstone formation. The white sand is loosely consolidated and when crushed consists largely of very fine grains.

Silica in the form of quartzite occurs 13 miles northwest of Crystal in the Arrow Canyon Mountains. The deposit is owned by Seth E. Evans of Goodsprings, Nev., and in 1936 it was under lease to W. M. Tiffany of Long Beach, Calif. while this deposit is not in the Muddy Mountains area, it is included here for convenience. Very little work has been done on the deposit and there is no information on any production.

The silica grains are tightly cemented and the exploitation of this deposit depends on using the material in lump form as ganister or for purposes requiring lump silica. Under the glass the silica appears virtually free from any impurities.

Bentonite Deposits

Small deposits of bentonite types of clay occur in the Muddy Mountains area. One of the deposits, covered by three unpatented lode claims owned by Robert A. Baker of Las Vegas, Nev., is northwest of Muddy Peak 9 miles southeast of Crystal. Development work consists of a shaft 30 feet deep and several shattered open-cuts. The bentonite occurs between limestone and sandstone, and the width varies from 6 to 14 feet. None has ever been produced.

Sodium Sulphate Deposit

A deposit of sodium sulphate owned by B. R. McKenna of Moapa, Nev., outcrops along the side of a prominent cliff in the White Basin about 3 miles south of the former American Borax Co. mine. It is accessible by poor automobile road from Crystal, Nev., a station on the Union Pacific R. R., 24 miles to the northwest. No attempt has been made to exploit the deposit. The sodium sulphate is intimately associated with siliceous impurities.

ST. THOMAS (VIRGIN RIVER) DISTRICT (Silica Sand, Magnesite, Gypsum, Salt, Sodium Sulphate)

The St. Thomas or Virgin River district comprises a large, vaguely defined area in the vicinity of St. Thomas, Nev. The town of St. Thomas has been abandoned, as eventually it will be flooded by the water impounded by Boulder Dam. The district is particularly rich in industrial mineral deposits, including silica sand; magnesite, gypsum, salt, and sodium sulphate; but due to their inaccessibility and distance from consuming centers, only silica sand and salt have been produced. The salt deposits have been worked mainly to supply local demand. The Los Angeles & Salt Lake

R.R., a part of the Union Pacific System, passes through Moapa, Nev., 14 miles northwest of Overton. From Moapa a branch railroad runs southeast to St. Thomas, a distance of 21 miles. About 5 miles of the southern part of this branch line will be abandoned when the water in Lake Mead reaches its maximum flow line.

The production of silica sand in Clark County is a comparatively recent industry, having been established less than 10 years. Production statistics are not available, but the total from the St. Thomas deposits is probably more than 200,000 tons. During the first quarter of 1937 the average monthly production was 3,500 tons, consisting of 2,500 tons of washed white sand and 1,000 tons of pink sand, the latter shipped without treatment. The bulk of the white sand wac produced by the Morledge and Veitch Co., formerly the Sparton Silica Co.

Both white and pink varieties of silica sand occur in the sandstone formations in the Valley of Fire northwest of St. Thomas. These sandstone beds of Jurassic age and of great thickness have been eroded into fantastic shapes that exhibit many hues of red, yellow, brown, pink, and white, which, in the bright light and heat of the desert, give the Valley of Fire its name.

In Clark County the production of a white glass sand suitable for the better grades of glassware depends upon a supply of water for washing to remove small amounts of clay, iron, and amorphous silica. Although extensive deposits occur in other parts of the County, it is doubtful whether they can produce a high-grade glass sand without washing. In the St. Thomas area, water for washing is available from the Muddy River, which joins the Virgin River a little below St. Thomas.

With the present import tariff of $1 per long ton on silica sand. the St. Thomas deposits compete with imported Belgian sand in the growing glass and foundry industries of southern California. The white sand, after washing, is suitable for the manufacture of fancy glassware and sodium silicate; the pink variety is shipped without treatment and is used for molding purposes and manufacture of the cheaper grades of colored glassware. Production of silica sand in Clark County is on a close-margin basis at the present time, with a chance to expand if industries requiring this material are established in the vicinity of Boulder Dam.

Morledge & Veitch Co.

The white silica-sand deposit owned by F. L. Morledge and Lloyd Veitch is covered by two placer locations of 40 acres each 3.8 miles southwest of the washing plant. The washing plant is on the St. Thomas branch of the Union Pacific R.R. 5 miles south of Overton, Nev. In March 1937, 25 men were employed in the production of about 2,500 tons of white sand per month. Production for the past 3 years, taken from the company records, has been as follows:

Year	Tons (2,000 lb.)
1934	16,180
1935	14,935 (closed 4 months)
1936	18,717

The deposit occurs in a small hill covered by capping of iron-stained and cemented detrital material varying in thickness from a few feet to 20 feet. Below the overburden the sandstone consists of rounded crystalline grains fairly uniform in size and loosely consolidated. The sandstone is firm enough so that explosives are necessary to reduce it to a size convenient for handling. However, a light blow will reduce the lump to fragments, and the sandstone disintegrates rapidly in water. The mine-run sand contains about 20 percent waste in the form of hand lumps, clay, and amorphous silica.

A representative analysis made by the Smith Emery Co. of Los Angeles, Calif., on the washed sand was as follows:

	Percent
Silica	99.290
Iron oxide	.051
Aluminum oxide	.220
Titanium oxide	.050
Calcium oxide	.090
Magnesium oxide	.020
Sodium and potassium oxide	.100
Loss on ignition	.180
Total	100.001

The tolerance for iron in silica sand is 0.05 percent when used for better grades of glass, and according to Morledge the finished sand can be maintained within this limit. Some of the sand is shipped for the manufacture of sodium silicate, and for this Purpose the iron content is not so important as it is in glass sand. The glass trade desires a product fairly uniform in grain size.

In the initial stages of mining, the overburden is first removed with a horse-drawn scraper, after which the sand is blasted and loaded directly into trucks by hand shoveling. Most of the sand mined in the past has been taken from an open-pit approximately 240 feet long, 80 feet wide, and 100 feet deep. From this pit the sandstone is broken to a convenient shoveling size with 12-percent Hercomite No. 6. Holes are drilled with hand augers. The sandstone is hand-shoveled into end-dump mine cars holding 1,500 pounds each and is trammed to a loading pocket at one end of the bit. A skip holding about 3/4-ton, traveling on rails, and operated by Continental hoist powered by 35-horsepower gasoline engine, is used to hoist the sand out of the pit. The skip dumps on a rail grizzly with 5-inch openings. The grizzly oversize contains some hard lumps of silica cemented with clay, and these are hand-sorted and rejected as waste. From the bin below the grizzly the sand is loaded into 5-ton trucks and hauled to the washing plant. The flow sheet of the washing plant is shown in figure 7.

The sandstone is crushed to 1/2-inch size by a set of rolls. It is disintegrated further by the action of water sprays in the trommel screen. Iron is present in the sand as nodules of ferruginous material resistant to disintegration, so that much of it is eliminated in the trommel over-size.

By successive washings in drag classifiers, the amorphous silica and clay are rejected. In the washing plant the coarser and finer sands are separated by hydraulic and drag classifiers. A screen test on the three sizes of sand produced was as follows:

Retained on (screen size)	20	40	60	80	100	-100
Molding sand	0.5	49.1	30.6	7.2	6.3	6.8
Glass sand	Trace	13.9	44.4	18.6	14.1	9.2
Sodium silicate sand	0	Trace	11.8	18.4	29.2	39.6

The finished product is dried in the air on a concrete storage platform. After it is dry it is loaded into railroad cars by a portable belt conveyor.

Power for the treatment plant is supplied by a 50-horsepower Western gasoline engine. The capacity of the plant is 10 tons of finished product per hour.

Nevada Silica Sand Co.

A deposit of white silica sand worked by the Nevada Silica Sand Co. is 4 miles west of Tokio Siding, a station on the St. Thomas branch of the Union Pacific R.R. several miles north of Overton, Nev. The deposit, which covers 160 acres, is worked under lease from the owner, Fay Perkins of Overton. The Nevada Silica Sand Co. is controlled by H. J. Stocker of Las Vegas, Nev., and associates.

The washing plant is 5 1/2 miles from the deposit, at a place 1 1/2 miles south of Tokio Siding, where water is available. Several years ago a pneumatic classification plant was erected at the deposit but it proved to be unsuccessful. In the latter part of 1936 eight men were engaged in producing silica sand. Judging from the extent of the open-pit working, the total amount of sand produced from this deposit has been approximately 15,000 tons.

The deposit is covered by an iron-stained sandstone capping whose depth ranges from a few feet to 10 feet. Below this capping the sand is fairly uniform in texture, loosely cemented, and in places stained with iron oxide. When the writer visited the property in November 1936 the workings comprised an open pit 100 feet long, 75 feet wide, and about 40 feet deep at the lower end. As depth is attained, the sandstone will probably contain less iron-stained material.

Mining is done with a 1/3-cubic-yard-capacity dragline scraper operated by a 35-horsepower, 4-cylinder, gasoline engine. The sand is broken with Hercomite No. 6 explosives. Hand-drilling is employed. The dragline transports the broken sandstone to an elevated bin equipped with rail grizzly having 5-inch openings. Some of the grizzly oversize is either iron-stained or too hard to be broken easily by hand-sledging, and this material is rejected by hand-sorting. The grizzly undersize drops into a 9-by 12-inch Blake-type crusher, which reduces it to a minus 1-inch product. From the bin the sand is loaded into either of two International trucks holding 8 and 4 tons, respectively, which haul it to the washing plant. The flow sheet of the washing plant is shown in figure 8.

From the mill bin, a belt-feeder delivers the sand to a trommel 3 1/2 feet in diameter and 8 feet in length equipped with a 14-mesh screen and water sprays. The trommel oversize is rejected as waste. The trommel undersize is discharged into a home-made simplex drag classifier. The classifier sand passes into a scrubber consisting of a wooden cylinder 5 feet in diameter and 18 feet long, which turns at a speed of 18 revolutions per minute. The scrubber discharges into a second home-made simplex drag classifier, which makes the finished product. The finished product is carried by belt conveyor to a draining platform, and after the sand is drained and dried in the open air it is loaded directly into railroad cars by belt loader.

The capacity of the washing plant is 60 tons of finished product in 9 hours with a crew of three men. Power for the plant is furnished by a 40-horsepower gasoline engine.

Other Silica-Sand Deposits

W. R. Cozart, of Overton, Nev., owns 800 acres of white silica sand several miles southwest of Overton. Only 112 carloads of this sand has been mined and was shipped in 1927-28. In 1937 this deposit was under lease to Paul and Fred Nunn, who were engaged in constructing a washing plant on the St. Thomas branch of the Union Pacific R.R. several miles south of Overton.

There is deposit of wind-blown pink sand (owned by U. V. Perkins of Overton, Nev.) 3 1/2 miles west of Logandale, Nev. The deposit is covered by two unpatented claims under lease to Morledge & Veitch Co., who were shipping about 500 tons of sand per month. This deposit consists of sand fairly uniform in groin size and with a silica content averaging about 97 percent.

The material is shoveled by hand into trucks and hauled 3 1/2 miles to a railroad spur for shipment.

In addition to the foregoing deposits, a number of small operators ship pink sand intermittently from dune deposits several miles west of Kaolin, Nev. The pink sand is used in the manufacture of colored glass- ware and for foundry purposes, and it is shipped without treatment. The deposits of pink sand are large and production is limited only by lack of market.

Magnesite Deposits

Extensive deposits of impure magnesite occur on the northeast side of the Muddy Mountains, 5 miles by road west of Kaolin, Nev., a station on the St. Thomas branch of the Union Pacific R.R. Magnesite claims were first located in 1915, although the deposits probably were known prior to that time because of their conspicuous outcrops.

The Success group of six unpatented lode claims, also covered by a placer location, is owned by Mrs. John F. Perkins, the Gentry estate, and others. The Prosperity group of five unpatented lode claims is owned by John F. Perkins and associates. The South group of three unpatented lode claims is held by Levi Syphus and associates. Most of the owners live at Overton, Nev.

The deposits have been prospected by a number of adits but no commercial shipments of magnesite have been made.

A detailed report of the geology of these magnesite deposits has been made by W. W. Rubey and Eugene Callaghan[27].

The magnesite occurs in white, fine-grained, sedimentary beds interbedded with white clayey dolomite and ranging in dip from 25° to 42°.

According to Rubey and Callaghan, the estimated tonnage of the magnesite in the Overton area is as follows:

Minimum grade of ore	Magnesite (short tons)			
	In beds 4 feet or more thick	In beds 2 feet or more thick	In beds 1 foot or more thick	In beds 6 inches or more thick
38 % of MgO	290,000	430,000	760,000	850,000
34 % percent of MgO	3,100,000	3,500,000	3,600,000	3,700,000
30 %of Mg0	4,500,000	4,900,000	5,000,000	5,100,000

These deposits have received attention as a source of magnesium in connection with the Power from Boulder Dam. When the reservoir (Lake Mead) formed by Boulder Dam reaches its maximum flow line, the magnesite deposits will be only a short distance from the shore of the lake. To be commercially valuable, the lime and silica impurities that occur in the magnesite would have to be removed by some-method of beneficiation. The magnesite is so fine-grained that physical methods of separating it from the impure constituents do not appear feasible.

Gypsum Deposits

There are large reserves of gypsum on the east side of the Virgin River, 10 miles below St. Thomas,.Nev. Although the fact that these deposits were there has been known for.many years, they have not been mined due to their inaccessibility and distance from market. However, when the water in Lake Mead reaches its maximum flow line the gypsum can easily be transported by barges to the Union Pacific R.R. at Boulder City. Very little work has been done on the deposits.

The Colorado River Exploration Co. owns nine patented claims com- prising 1,440 acres. Another group of 11 unpatented claims of 160 acres each is owned by C. C. McDonald, of Overton, Nev., and associates.

Salt Deposits

Remarkable deposits of rock salt are exposed on the cliffs along the west side of the Virgin River for a distance of 10 miles south from a point a mile below St. Thomas, Nev. These deposits have been acquired by the Federal government, since they will eventually be submerged as the water rises in the reservoir behind Boulder Dam.

The salt beds were mined by a tribe of Indians living in the Virgin River Valley about 1,500 years ago. Evidence of this ancient mining activity may be seen on the walls or floors of a natural cave in one of the salt mountains 4 miles below St. Thomas. The Indians mined the salt by chipping circular channels with stone hammers and prying off the slabs. A number of stone hammers found in the cave by archaeologists may be seen in the Lost City museum, a short distance south of Overton, Nev.

White men mined salt in this locality as far back as 1866 in order to supply salt for the chlorination mills in the Mineral Park district, Ariz., and the Eldorado Canyon district, Nev. Although the salt deposits are virtually inexhaustible, little has been produced because of transportation difficulties and distance from large markets in competition with other sources of supply.

About 1932 some rock salt was shipped to Los Angeles by a company called the Virgin River Salt Co. Judging from the old workings, no large amount of salt was produced. For many years an average of about 200 tons of salt has been mined annually to satisfy local demands for stock purposes. This salt is very desirable for stock use because it weathers slowly.

In February 1937 the Civilian Conservation Corps at Moapa, Nev., was engaged in hand-mining the salt for stock use. Several thousand tons of this salt will be mined and stored before the deposits are flooded.

The salt bed of probably greatest commercial value is 4 miles south of St. Thomas and is about 100 feet thick. It is covered with a coating of sand and silt, but under this coating it is hard and solid and in order to be mined must bc blasted. Some specimens of the salt are nearly as transparent as plate glass, but the bulk of it is mixed with clay and silt.

Sodium Sulphate (Glauberite) Deposit

Sodium sulphate (glauberite) occurs on the west side of the Virgin River about 5 miles south of St. Thomas. This deposit is on patented land formerly owned by the Stewart brothers. This land has been purchased by the Federal Government, since it will be submerged when water rises to the crest of Boulder Dam. The deposit has been explored by an adit 200 feet long, near the end of which is a winze 40 feet deep. No production has ever been made.

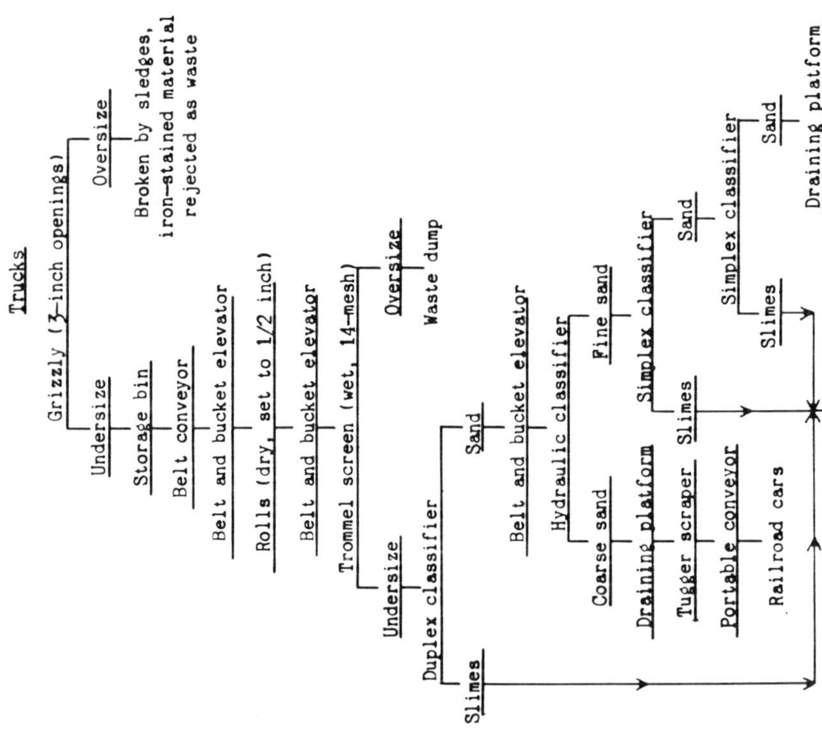

Figure 7.- Flow sheet of Morledge-Veitch silica-sand washing plant near Overton, Clark County, Nev.

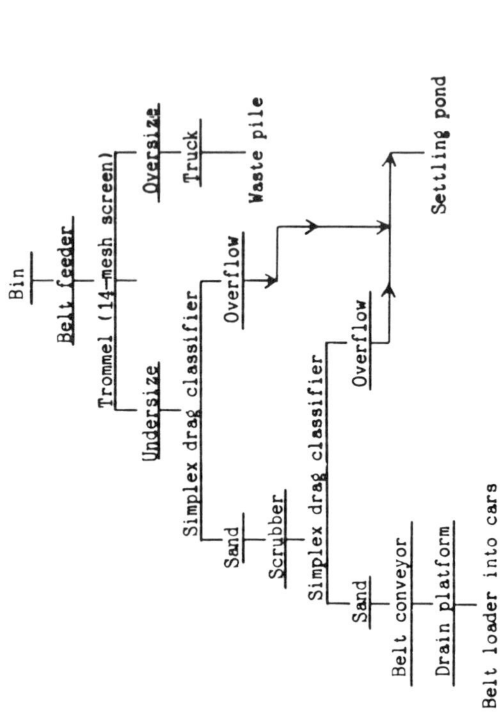

Figure 8.- Flow sheet of the Nevada Silica Sand Co. washing plant near Tokio Siding, Clark County, Nev.

The sodium sulphate occurs as a crystalline mass of glauberite in horizontal sedimentary beds. Composite samples of the material analyzed 46.8 percent sodium sulphate, 45.85 percent calcium sulphate, and 7.35 impurities, including sodium chloride, iron, alumina, and silica. Glauberite when pure is composed of 51.1 percent sodium sulphate and 48.9 percent calcium sulphate, and the foregoing samples indicate that the deposit is fairly pure. The reserves are probably large.

SEARCHLIGHT DISTRICT
(Gold, Silver)

The Searchlight district is in southeastern Clark County, 40 miles south of Boulder City, Nev., and 22 miles east of Nipton, Calif. Nipton, on the main line of the Union Pacific R.R. is the nearest shipping point. The town of Searchlight, with a population of about 200, is 16 miles west of the Colorado River at an elevation of 3,600 feet.

The first claims in this area were located in 1897 by G. F. Colton, John Swickard, and others. The first important discovery was made on the Searchlight claim, now part of the Duplex mine, and the Quartette mine was discovered shortly after. The latter was financed by Col. C. Hopkins of Buffalo, N. Y. In 1902 the Quartette Mining Co. erected a 20-stamp mill on the Colorado River, 16 miles east, and built a narrow-gage railroad to connect the mine and the mill. In 1903 water was encountered in the Quartette shaft and a second mill of 20-stamp capacity was built at Searchlight.

The greatest activity occurred from 1902 to 1916, during which period a large number of companies were organized, many of which were merely stock- selling ventures. From 1916 to the present time mining has been principally in the hands of lessees. In 1936 an average of 60 men were employed on various properties in the district.

The mineralized area in the vicinity of Searchlight is 3 miles long and 1 mile wide. The ore occurs in fissure veins near a major fault contact between quartz monzonite and Tertiary eruptives. The fissures contain very little quartz, and the pay ore usually is a mass of shattered country rock or gouge material stained with iron oxides and carrying free gold associated with copper and lead minerals generally oxidized. Production of the district is shown in table 4.

A number of properties at Camp Thurman, Camp Dupont, Black Mountain, and Newberry Mountain are included in this report under the Searchlight district.

Custom Milling

The Kelsey custom mill at Searchlight, owned by W. H. Kelsey, was built in the latter part of 1935 to treat custom ores derived from various properties within the Searchlight area. The capacity of the mill is 30 tons per 24 hours, but the supply of ore has been insufficient to permit steady operation of the mill. In 1936 an average of 400 tons or ore. valued at from $7 to $100 per ton, were treated per month.

The flow sheet of the mill is shown in figure 9. A crew of four men is required to operate the mill on a one-shift-per-day basis. The cost of the plant, with all used equipment excepting the stamps, which were new, but including laboratory and assay office, was $20,000. Power for milling is furnished by a 100-horsepower, Fairbanks-Morse, Y-type, 3-cylinder, Diesel engine. Water for milling is pumped from the Santa Fe shaft near the mill. Custom milling charge is $4 per ton, and payment is made for 90 percent of the assay value of the gold estimated at $33 per ounce and 90 percent of the assay value of the silver less one ounce at 70 cents per ounce.

Blossom Mine

The Blossom mine, comprising five patented claims, is owned by William H. Crozier of Los Angeles, Calif. This mine, originally part of the Southern Nevada property, was purchased by Crozier at a tax sale several years ago. In 1936 the western portion of the mine was leased to A. S. Gaines and Chris Kirkeby of Searchlight, Nev. During the last 6 months of 1936 the lessees had shipped 27 carloads of ore averaging about $100 per ton. An average of 12 men are employed. In the early days the mine is reported to have produced $325,000.

Development work on the Blossom property comprises several shafts, the deepest of which is 430 feet, and other workings, totaling in all about 2,000 feet. Mining equipment on the Gaines-Kirkeby lease includes a Fairbanks-Morse geared hoist driven by 6-horsepower gasoline engine and blacksmith shop.

The formation is andesite and quartz monzonite. In former years the production of the Blossom mine was derived from a blanket vein averaging 3 feet in width and outcropping around the sides of a small hill. This form of deposit is unique in the district, as all the other deposits are inclined veins of the fissure type. On the Gaines-Kirkeby lease the vein varies in width from a few inches to 15 inches. and the values occur in crushed andesite and gouge. Most of the ore has been mined between the surface and the 110-foot level. Values are chiefly in gold and silver.

Because of the erratic distribution of values the ore is mined by irregular workings with little regard for development. There are no pronounced structural features to distinguish the vein material, and constant panning is necessary to distinguish between ore and waste. In places where the ore is continuous the resuing method of mining is employed. In February 1937 the inclined shaft had reached a depth of 160 feet. The ore is hoisted in a bucket (600 pounds capacity) sliding on skids. The ground is soft and drilling is done with hand augers or hand steel.

A shipment of ore made by A. S. Gaines and Chris Kirkeby to the U. S. Smelting & Refining Co. on November 20, 1936, furnished the following information:

Metal quotation:	Au at	$34.9125	
	Ag at	.77	
		Oz	
Settlement assay:	Au	3.845	
	Ag	47.62	
		Percent	
	Insol.	83.4	
	Fe	3.2	
	S	.2	
	Lime	1.0	
Metal payment	100 percent Au at $32.3183		$124.26
	Ag at .77		36.67
	Total metal value		160.95
	Less 5 percent value of Ag		1.83
	Gross metal value		159.10
	Less treatment		7.70
	Net value per ton		151.40

TABLE 4. - Gold, silver, copper, and lead production from Searchlight District, Clark County, Nevada, 1904-1935

(Compiled by Charles White Merrill, supervising engineer, San Francisco office, Mineral Production and Economics Division, U. S. Bureau of Mines)

Year	Placer						Lode			
	No. of mines	Gold		Silver		Total value	No. of mines	Ore, short tons	Gold	
		Fine ounces	Value	Fine ounces	Value				Fine ounces	Value
1904	------	------	------	------	------	------	3	16,750	18,401.17	$380,386
1905	------	------	------	------	------	------	6	38,069	19,329.44	399,575
1906	------	------	------	------	------	------	5	45,668	25,144.60	519,785
1907	------	------	------	------	------	------	6	45,921	23,440.98	484,568
1908	------	------	------	------	------	------	7	52,193	13,137.34	271,573
1909	------	------	------	------	------	------	8	68,931	16,236.83	335,645
1910	------	------	------	------	------	------	13	27,331	10,405.32	215,097
1911	------	------	------	------	------	------	10	1,850	966.48	19,979
1912	------	------	------	------	------	------	9	4,153	2,237.05	46,244
1913	------	------	------	------	------	------	15	5,989	6,653.84	137,547
1914	------	------	------	------	------	------	5	3,057	3,998.53	82,657
1915	------	------	------	------	------	------	13	7,766	4,984.66	103,042
1916	------	------	------	------	------	------	15	6,322	3,902.75	80,677
1917	------	------	------	------	------	------	14	12,866	3,444.93	71,213
1918	------	------	------	------	------	------	14	1,700	2,649.06	54,761
1919	------	------	------	------	------	------	9	6,980	2,854.37	59,005
1920	------	------	------	------	------	------	7	2,131	3,432.35	70,953
1921	------	------	------	------	------	------	9	1,182	3,600.65	74,432
1922	------	------	------	------	------	------	7	1,441	1,661.29	34,342
1923	------	------	------	------	------	------	6	1,142	1,006.93	20,815
1924	------	------	------	------	------	------	2	850	830.12	17,160
1925	------	------	------	------	------	------	5	2,833	3,313.93	68,505
1926	------	------	------	------	------	------	4	1,881	1,879.42	38,851
1927	------	------	------	------	------	------	4	106	334.90	6,923
1928	------	------	------	------	------	------	3	77	223.78	4,626
1929	------	------	------	------	------	------	3	184	105.26	2,176
1930	------	------	------	------	------	------	3	124	19.40	401
1931	------	------	------	------	------	------	8	20,855	3,276.29	67,727
1932	------	------	------	------	------	------	7	2,973	2,018.41	41,724
1933	------	------	------	------	------	------	10	2,996	912.58	23,326
1934	1	8.32	$ 291	------	------	$ 291	24	6,701	1,745.52	61,006
1935	1	2.96	10,656	------	------	10,656	35	29,617	3,825.62	133,897
Total	2	11.28	10,947	------	------	10,947	------	420,644	185,973.80	3,928,618

5883

- 70 -

TABLE 4. - Gold, silver, copper, and lead production from Searchlight District, Clark County, Nevada, 1904-1935 (Continued)

(Compiled by Charles White Merrill, supervising engineer, San Francisco office, Mineral Production and Economics Division, U. S. Bureau of Mines)

Year	Lode							Average recoverable value of ore per ton[1]	Total value (lode and placer)
	Silver		Copper		Lead		Total value		
	Fine ounces	Value	Pounds	Value	Pounds	Value			
1904	13,245	$ 7,682	------	------	------	------	$ 388,068	23.17	$ 388,068
1905	28,528	17,402	22,808	$ 3,558	12,064	$ 567	421,102	11.06	421,102
1906	11,543	7,849	11,182	2,158	9,655	550	530,342	11.61	530,342
1907	7,494	4,946	37,063	7,413	38,113	2,020	498,947	10.87	498,947
1908	10,917	5,736	14,954	1,974	44,857	1,884	281,217	5.39	281,217
1909	13,544	7,043	22,916	2,979	36,209	1,557	347,224	5.04	347,224
1910	11,489	6,204	93,848	11,919	45,847	2,017	235,237	8.61	235,237
1911	2,136	1,132	12,095	1,512	7,659	345	22,968	12.42	22,968
1912	2,562	1,576	10,126	1,671	10,032	451	49,942	12.01	49,942
1913	5,903	3,565	53,971	8,365	61,006	2,684	152,161	25.41	152,161
1914	2,855	1,579	5,556	739	21,919	855	85,830	28.08	85,830
1915	3,546	1,798	20,970	3,670	35,562	1,671	110,181	14.19	110,181
1916	4,340	2,856	24,392	6,000	9,951	687	90,220	14.27	90,220
1917	9,073	7,476	98,923	27,006	59,453	5,113	110,808	8.61	110,808
1918	8,395	8,395	44,477	10,986	77,863	5,528	79,670	46.86	79,670
1919	5,502	6,162	18,074	3,362	39,307	2,083	70,612	10.12	70,612
1920	5,275	5,750	44,941	8,269	91,436	7,315	92,287	43.31	92,287
1921	9,877	9,877	31,128	4,016	98,048	4,412	92,737	78.46	92,737
1922	5,024	5,024	12,847	1,734	44,375	2,441	43,541	30.22	43,541
1923	2,708	2,221	22,150	3,256	47,651	3,336	29,628	25.94	29,628
1924	1,267	849	5,801	760	46,050	3,684	22,453	26.42	22,453
1925	2,586	1,795	14,974	2,126	86,768	7,549	79,975	28.23	79,975
1926	1,720	1,073	7,430	1,040	81,977	6,558	47,522	25.26	47,522
1927	306	174	1,429	187	10,415	656	7,940	74.91	7,940
1928	252	147	1,929	278	8,991	521	5,572	72.36	5,572
1929	58	31	654	115	2,743	173	2,495	13.56	2,495
1930	27	10	------	------	------	------	411	3.31	411
1931	5,826	1,690	6,022	548	274,687	10,163	80,128	3.84	80,128
1932	4,853	1,369	4,353	274	355,458	10,664	54,031	18.17	54,031
1933	1,182	414	2,261	145	12,040	445	24,330	8.12	24,330
1934	8,478	5,477	2,237	179	4,951	183	66,845	9.98	67,136
1935	12,317	8,853	9,195	763	72,271	2,891	146,404	4.94	157,060
Total	202,828	136,205	658,706	117,002	1,747,358	89,003	4,270,828	10.15	4,281,775

1/ Not to be confused with average assay value.

	Lb.		
Wet weight	122,560		
Less 6.7 percent moisture	8,212		
Net weight	114,348 or 57.174 tons at $151.40		$8,656.14
Deductions:	Freight, $11.50 per ton	$704.40	
	Assay charges	6.00	
	Sampling charges	44.90	
		755.62	755.62
	Net proceeds		7,900.52

The cost of hauling to Nipton. Calif., on the Union Pacific R.R., 22 miles west of the mine, is $3 per ton on contract. The royalty payments, based on the net smelter returns, are as follows:

Value of ore	Percent royalty
$20 or less	10
$20 to $45	15
$45 to $65	20
$65 or more	25

Quartette Mine

The Quartette mine, owned by Mrs. Frank Miller of Los Angeles, Calif., has been the best producer in the Searchlight district, with a production in excess of $2,000,000. The period of greatest activity was from 1905 to 1912, when this mine was operated by the Quartette Mining Co. In recent years the property has been worked intermittently by lessees. In 1936 the Mollin Mining Co., controlled by Rex Moss of Los Angeles, Calif., erected a cyanide plant to treat the Quartette tailings. This plant superseded a flotation mill on the Quartette property.

The mine is developed by three shafts 500, 900, and 1,350 feet deep and extensive underground workings. The maximum vertical depth attained in the mine workings is 900 feet. Most of the mining equipment has been removed.

According to Ransome[28], the Quartette lode is a strong fissure zone striking about north 65° west and dipping from 20, to 70° to the south. The main ore shoot was 350 feet long and up to 50 feet in width. Formation is andesite and gneiss cut by aplite dikes. The ore consists of crushed country rock generally oxidized and carrying values chiefly in gold, with some silver associated with chrysocolla, cuprite, iron oxides, and quartz. The less frequent constituents in the ore are cerussite, wulfenite, cuprodescloizite, azurite, malachite, and vanadinite.

The flow sheet of the cyanide plant for the treatment of tailings is shown in figure 10. The all-slime cyanidation process with countercurrent decantation is employed. Capacity of the plant is 85 tons per day, and it grinds from about 50 percent plus 48-minus S-mesh to minus 100-mesh. Cyanide at the rate of 1 pound and lime at the rate of 2 pounds per ton of solution are added to the ball mill. Precipitation is effected with zinc shavings in zinc boxes made from 50-gallon oil drums. The cyanide strength for precipitation is 1 1/2 pounds per ton of solution. To improve preciptation, lead acetate solution is added at the head of the zinc boxes. Power for milling is furnished by a 300-horsepower Fairbanks-Morse Diesel engine. Water is pumped from several shafts in the vicinity.

The Duplex mine dumps have been sampled by the Mollin Mining Co. for mill material, and it is reported that by screening the dumps through 3/4-inch mesh screen about 40,000 tons of material averaging about $6 per ton can be obtained.

Chief of the Hills Mine

The Chief of the Hills mine, owned by the Riverview-Cumberland Mining Corporation (Emil H. Koehl, president, Hollywood, Calif.), comprises a group of six claims [four patented) 3.1/2 miles southeast of Searchlight. In 1936 the property was leased to J. T. Woodward and associates for 10 years. From July 1936 to February 1937 Woodward, working with three men, shipped 175 tons of ore, averaging $11.77 per ton, to the Kelsey mill at Searchlight. Total production has been about $100,000.

Property is developed by several adits; the main adit is 1,100 feet long and from this adit a winze has been sunk to a depth of 165 feet. Two levels have been driven from the winze below the adit level. Total workings comprise about 3,500 feet. Mining equipment includes a Chicago pneumatic 2-stage compressor (capacity 300 cubic feet), a 6-horsepower geared hoist driven by gasoline engine, a LeRoi Rix portable compressor, blacksmith shop, assay office, and mining tools. The mill on the property is equipped with a Dodge crusher (12 by 4 inches), a 25-ton, slow-speed, Lane Chilean mill, and two amalgamation plates (5 feet wide and 10 feet long). The power plant for the mill has been removed.

The ore occurs in a shear zone in andesite porphyry. The shear zone ranges in width from 1 to 6 feet and dips about 70°. It has been disturbed considerably by faulting. Values are chiefly in gold, with a small amount of silver.

Duplex Mining Co.

The Duplex Mining Co., controlled by Mrs. Frances Roller of Searchlight, Nev., comprises seven patented claims about 1/2 mile north of the Quartette mine. This property has been the second largest producer in the Searchlight district, with a production of about $1,000,000, most of which was made during the early days of the camp. In recent years the property has been worked intermittently by lessees. About 4 years ago an attempt was made to work the mine on a large scale, but this venture was unsuccessful. In 1936 lessees are reported to have shipped about 1,000 tons of ore to the Kelsey custom mill at Searchlight.

The mine is developed by several shafts, the deepest of which is 700 feet, and extensive underground workings.

The ore occurs in several veins ranging from 2 1/2 to 5 feet in width. The formation is principally monzonite. Values are in gold and silver, which at depth is associated with copper, lead, and zinc. The ore mined near the surface was treated by amalgamation, but at depth the ore became base, so that it was either shipped to smelters or treated in a flotation mill at the mine.

Southern Nevada Group

The Southern Nevada group of two patented claims, owned by Mrs. Ada K. Becker of Los Angeles, Calif.,.and Mrs. M. L. Reynolds of Buffalo, N. Y., is about 1 mile north of Searchlight, Nev. In 1936 the property was leased to J. A. Leavitte and Ernest Sandquist of Searchlight on a royalty basis of 10 percent of the net returns. when the writer visited the property the two lessees were screening and hand-sorting gob material from the old stopes.

Property is developed by several shafts, the deepest of which is 400 feet, and lateral workings, totaling about 4,000 feet. Considerable water was encountered at about 300 feet. Equipment includes a geared hoist driven by 5-horsepower gasoline engine.

Two veins occur on the property, one vertical and the other dipping about 30°, and averaging about 1 1/2 feet in width. Formation is andesite. Values are in gold, with some silver.

The stope filling is screened through a 1/2-inch mesh screen, and the undersize product is sent to the Pompeii mill owned by John R. Mendenhall. The oversize is hand-sorted. About 50 percent of the material is rejected. The screened product averages $12 per ton. The milling charge at the Pompeii mill is $4.50 per ton, and.payment is made on 90 percent of the assay value of the ore. Trucking to the mill, 1 1/2 miles distant, costs 75 cents per ton, and the lessees were required to do the loading by hand shoveling. The lessees averaged better than wages.

Pompeii Group

The Pompeii group, comprising eight patented and two unpatented claims owned by John R. Mendenhall, lies 2 1/2 miles north of Searchlight. A small tonnage of ore was mined from this property in the early days. Development comprises a shaft 330 feet deep, inclined 60°, and about 1,000 feet of lateral workings. In sinking the shaft a large flow of water was encountered at a depth of 265 feet. Equipment includes a 25-horsepower, Fairbanks-Morse, single-drum, geared hoist and a 50-horsepower, Fairbanks- Morse, gasoline engine gear-connected to a Cornish pump in the shaft. Mining is done by hand methods. The mill on the property is equipped with a jaw crusher (6 by 9 inches), 10 stamps (850 pounds each), 2 Challenge feeders, 2 amalgamation plates (each 5 feet wide by 10 feet long), and a 40-horsepower, Fairbanks-Morse, gasoline engine.

The main vein strikes northwest-southeast and dips southwest about 50°. the ore occurs as lenticular bodies in brecciated andesite. The best values are found where the vein is traversed with seams of quarts and cal- cite. the demarcation between ore and waste is usually poorly defined. Values are in gold and silver. when the writer visited the property two men were employed in prospecting the old workings near the surface.

Black Mountain Mining Co.

The Black Mountain Mining Co., owned by Ad Rickard of Searchlight. Nev., and associates, comprises 14 unpatented claims on Black Mountain 12 miles east of Searchlight and 4 miles west of the Colorado River. No production has ever been made. Development comprises an adit 250 feet long and scattered surface workings; totaling in all about 300 feet. Mining is done by hand methods.

Free gold occurs in a stringer zone 6 to 12 feet wide in hard andesite. The adit has been driven on the strike of this zone and, according to Rickard, car samples for the first 75 feet assayed from $5 to $6 per ton.

Goldenrod Group

The Goldenrod group of five patented claims (formerly known as the Lloyd Searchlight property) is owned by John Nelson of Searchlight, Nev., and associates. It lies 16 miles by road southeast of Searchlight in an area sometimes called the Thurman district, after John Thurman who made the original discovery in this locality in 1906. In February 1937 two men were employed on the property. In 1936, 50 tons of ore, averaging $70 per ton, were shipped. In former years a small amount of ore was mined and treated locally in a small amalgamation mill equipped with two Nissen stamps.

Development comprises a shaft 200 feet deep, another 75 feet deep, and about 1,000 feet of lateral workings. Equipment includes a Gardner Rix compressor driven by 6-cylinder Buick engine, a small hoist geared to a 3 1/2-horsepower gasoline engine, a blacksmith shop, and mining tools.

The ore occurs in a quartz vein that strikes about north 60° east, dips about 80°, and is 6 to 18 inches wide. The country rock is of granitic composition. Values are principally in free gold.

Jackdaw Group

The Jackdaw group of three unpatented claims (formerly known as the Old Roman mine), owned by J. J. McDonald of Searchlight, Nev., lies on the north side of Newberry Mountain, 18 miles southeast of Searchlight, Nev. It has been worked intermittently on a small scale for many years. Production has been probably about $15,000 in shipping ore. In February 1937 one man was working under lease agreement.

Development comprises two adits and subsidiary workings, totaling in all about 1,500 feet. Equipment includes an Ingersoll-Rand compressor (6 by 8 inches), belt-driven by a Mack truck engine. A small cyanide leaching plant on the property is equipped with a Cottrell ball mill (3 1/2 by 5 feet), jaw crusher (7 by 9 inches), and four concrete leaching tanks holding about 15 tons each. Power for the mill is furnished by a Ford automobile engine. Only a small amount of ore has been treated in this mill.

The ore occurs in a quartz fissure-vein. The dip is 60°, and the average width is about 2 feet. Values are in silver, associated in places with malachite and azurite. Formation is granite.

Homestake Group

The Homestake group of seven unpatented.claims, owned by J. J. McDonald of Searchlight, Nev., lies in the southeast corner of Clark County, 9 miles north of Hiko Springs and 39 miles by road southeast of Searchlight. This mine was discovered by soldiers from Fort Mohave, Ariz., in the early sixties, and the remains of three steam plants on the ground indicate that it was exploited very early. In the early days the mine was equipped with a 20-stamp amalgamation mill. According to McDonald, production has been about $150,000. Since 1910 McDonald has worked the property intermittently on a small scale, employing one to three men. Selected ore averaging $40 per ton is hauled to Cottonwood Camp 18 miles north, where it is treated in a small mill that employs amalgamation and cyanidation.

Development comprises a shaft 300 feet deep, inclined 66°, and several adits, which, with other workings, total about 4,000 feet. There is a small gasoline hoist at the shaft. Mining is done by hand methods.

The ore is in three parallel fissure veins that strike east and west and dip about 50° to the north. Formation is pre-Cambrian granite. The main vein, on which most of the work has been done, is from 4 to 20 feet wide, averaging 12 feet, and it can be traced on the surface for several thousand feet. Values are chiefly in gold, with small amounts of silver, in a gangue of iron-stained quartz.

OTHER PROPERTIES

Searchlight Mine

The Searchlight mine, comprising a group of 11 claims, owned by George W. Lichtenberger of Los Angeles, Calif., lies 1 mile north of Searchlight, Nev. It was formerly operated by the Searchlight Mine & Milling Co. In 1936 the mine was leased to J. L. Harper of Los Angeles, and in December a crew of 10 men was employed on development work. Production has been about $400,000. Property is developed by a shaft 340 feet deep and several thousand feet of underground workings. Mining operations in former years were handicapped by a large flow of water. Equipment installed by the present operators includes an amalgamation-concentration mill having a capacity of 18 tons per day, a Worthington portable compressor, and a small gasoline hoist.

St. Louis Group

The St. Louis group of eight unpatented claims, owned by W. H. Barton and associates, lies 10 miles northerly of Searchlight, Nev. In 1936 lessees produced ore valued at $22,000, which was treated at the Kelsey mill.

Oakland Group

The Oakland group of four unpatented claims; owned by Albert C. Calkins, lies 3 miles west of Searchlight. In 1936 property was under bond and lease to William G. Smith and associates and three men were employed in the latter part of the year to sink a shaft to intersect a vein exposed on the surface. Development comprises several shallow shafts, short tunnels, and open cuts; totaling about 500 feet. Equipment includes a 25-horsepower Western geared hoist, an Ingersoll-Rand Imperial No. 11 2-drill compressor, and a small amalgamation mill. The mill contains a jaw crusher (7 by 10 inches) and four Nissen stamps driven by a Dodge automobile engine, and two amalgamation plates (3 feet wide and 8 feet long). The ore occurs in fissure veins in andesite near a contact with monzonite. In sinking the shaft a flow of water, estimated at about 2,500 gallons per day, was encountered at a depth of 80 feet.

Bay City Group

The Bay City group of three unpatented claims, also owned by Albert C. Calkins adjoins the Oakland group on the south. This property is under lease to E. C. Douglas, who produced ore valued at $1,000 in 1936. The ore was treated in the mill on the Oakland property. Development includes a shallow shaft and other workings totaling about 200 feet. Mining equipment includes a small gasoline hoist.

Camp Dupont

In the vicinity of Camp Dupont 17 miles northeast of Searchlight, a number of claims are owned by individuals from Las Vegas, Nev., and Searchlight. Although considerable work has been done, all the properties are in the prospect stage. In 1922 the Dupont Copper Mines Co. (probably defunct) acquired the Sazarac, Bornite, and other groups of claims in this area, but no production was ever made.

In February 1937 the only activity was on the Big Shot group of two unpatented claims owned by George H. Kent of Searchlight and C. G. Duff of Las Vegas, Nev. About 35 tons of ore were mined from this property in 1936. The ore was treated in a two-stamp amalgamation mill on the Colorado River 6 miles to the east. Development work comprises an adit 150 feet long driven on a narrow vein. Mining is done by hand methods. The vein ranges in width from a few inches to 2 feet, strikes east and west and dips 60, to the south. Formation is rhyolite. Values are in gold in a gangue of quartz and crushed country rock stained with iron oxides.

SLOAN DISTRICT
(Limestone, Dolomite, Radium)

The Sloan district, in the vicinity of Sloan, a station on the main line of the Union Pacific R.R. connecting Salt Lake City, Utah, and Los Angeles, Calif., is about 20 miles a little west of south from Las Vegas, Nev., via Highway 91. The only production from this area has been limestone and dolomite. The occurrences of carnotite have proved to be too small to have commercial interest as a source of radium.

The Sloan limestone deposit, discovered in 1910, was first worked in 1912 to supply lime for the treatment of gold and silver.ores at Goldfield, Tonopah, and other Places where the cyanide process was employed. The deposit was mined by the Nevada Rock Corporation up to 1927, when it was purchased by the U. S. Lime Products Corporation, the present owners. In addition to mining limestone, this company started to mine the dolomite that occurs above the limestone bed. The total production of limestone and dolomite from this deposit is probably more than a million and a half tons. This operation is interesting not only because of the magnitude of the deposit but because it is in the desert region far removed from the centers of population where lime is used. Los Angeles, the principal consuming center, is 280 miles from Sloan.

United States Lime Products Corporation

The U. S. Lime Products Corporation employs an average of about 50 men in the annual production of about 50,000 tons of crude limestone and dolomite and about 25,000 tons of lime and dolomite products. At Sloan the company operates a plant for converting the dolomite and limestone into lime, lime-hydrate, dolomitic lime, and dolomitic lime-hydrate. Crude limestone is shipped to southern California for use in the refining of sugar and in open-hearth steel furnaces.

Plant equipment includes a Blake-type crusher (9 by 38 inches), two rotary kilns (7 and 8 feet, respectively, in diameter and 100 feet in length), four vertical kilns, hydrators, steel silos, trommels, elevators, bagging machines, and other equipment for a complete plant. Power for the plant is furnished by Diesel engines generating electricity. The company also maintains a control laboratory at Sloan for making chemical and physical tests on the lime products in order to maintain standards of quality.

The limestone and dolomite beds form an isolated mountain having a maximum elevation of 1,000 feet above the surrounding terrain. The limestone bed is about 150 feet thick and lies nearly horizontal. The overlying dolomite has an average thickness of about 600 feet. Both the limestone and dolomite are dense and uniform in composition so that selective mining is not necessary. The composition of the limestone and dolomite averages as follows:

Limestone	Percent
$MgCO_3$	1.72
SiO2	.72
Fe_2O_3; Al_2O_3	.45
$CaCO_3$	97.11
Total	100.00

Dolomite	Percent
$MgCO_3$	44.35
SiO_2	.14
Fe_2O_3; $A_{l2}O_3$	.37
$CaCO_3$	56.14
Total	100.00

The limestone is mined by quarrying and shrinkage stoping. Originally the limestone bed formed a prominent cliff so that it could be mined by quarrying with little dilution from the overlying dolomite. In recent years. however, the amount of overlying dolomite quarried with the limestone became excessive and preparations were made to mine the limestone by shrinkage stoping. For this purpose development headings 12 feet and 8 feet in width are driven at the base of the limestone bed. The shrinkage stopes are about 30 feet wide, 80 feet high, and up to 250 feet long. Pillars left between successive stopes average 20 feet in width. To facilitate drawing off the broken ore from the stopes, drifts are driven into the pillars, and from these drifts small crosscuts tap the stopes at convenient intervals. At the time of writer's visit the shrinkage-stoping operations were in the experimental stage, and all the details had not been worked out completely.

The stopes are advanced by drilling on broken rock. Holes are drilled from 8 to 12 feet deep with jackhammers and 7/8-inch hollow hexagonal steel with detachable bits. Blasting is done with 40-percent gelatin dynamite and No. 6 detonators.

The broken limestone is shoveled off the floor of the level by hand into 1-ton mine cars. Eventually, mucking machines may be employed. Tramming is done by hand. The crushing and lime-burning plants are below the quarry floor, so that the limestone is transported by gravity trams of the 3-rail type.

Carnotite Deposit

In 1915 carnotite was discovered in a railroad cut several miles south of Sloan by N. E. Williams of Las Vegas, Nev.

The carnotite, associated with calcite and manganese oxides, occurs as thin coatings along joint planes in rhyolite. Very little prospecting work has been done, and the deposits are too small to be of commercial interest.

SUNSET DISTRICT
(Gold, Feldspar)

The Sunset district is in southern Clark County near the California boundary line 4 1/2 miles east of Desert, Calif., a station on the Los Angeles & Salt Lake R.R. This district adjoins the Crescent district on the north. The only productive property in this area has been the Lucy Gray mine, discovered by Lee Bright in 1905. This property is said to have produced about $50,000 and, judging from the tailings pile at the mine, comprising about 4,000 tons, this is probably substantially correct. The property formerly was equipped with a Lane slow-speed Chilean mill and cyanide leaching equipment.

Deposits of feldspar in pegmatite dikes were located in this area in 1936, but no production has been made.

Lucy Gray Mine

The Lucy Gray mine comprises nine unpatented claims owned by the Bright brothers, John and Will, of Los Angeles, Calif. The property was last worked under company management in 1920, but since that date lessees have made sporadic shipments. In February 1937 the property was under 10-year lease to J. W. Collis and J. H. Roese of Los Angeles, who were pre- paring to equip it. The lease specifies royalty payments of 10 percent on the net mill or smelter returns.

Development includes a two-compartment vertical shaft 300 feet deep with main levels at 100-, 200-, and 300-foot elevations, and five short adits none of which is over 100 feet long. Total workings comprise 3,200 feet. Most of the equipment has been removed from the property.

The ore deposits occur as pockets in shattered zones in a granitic gneiss formation. The values are principally in gold, with minor amounts of silver. The gangue is chiefly jaspery quartz and iron oxides.

Feldspar Deposits

Feldspar occurs in the Sunset district 4 miles east from Desert, Calif. The deposits are covered by three unpatented claims located by George Noltkamper of Nipton, Calif., in 1936. Very little work has been done on the claims.

The feldspar occurs in numerous pegmatite dikes varying from a few feet to 30 feet in width. In most of the dikes the feldspar is intimately associated with quartz and biotite. However, in one dike averaging 30 feet in width and outcropping for a length of 500 feet masses of fairly clean feldspar are exposed.

FOOTNOTES

1 Statutes of the State of Nevada, 24th Session, 1901, p.12.

2 The population of incorporated Las Vegas in 1930 was 5,952.

3 The resulting lake covered much of lower Moapa Valley and submerged the community of St. Thomas.

4 Report of the Colorado River Commission of Nevada to Gov. Richard Kirman and the Nevada State Legislature, period Sept. 1, 1935, to Jan. 1, 1937, Alfred Merritt Smith, Secy.

5 Stewart, Helen J., History of Nevada: Edited by Samuel P. Davis, vol. 1 1913, pp. 217-218.

6 Gale, H. S., in Phalen, W. C., Potash Salts: U. S. Geol. Survey Mineral Resources, 1915, pt. 2, 1917, pp. 111-112.

7 Bradley. W. G., Methods and Costs of Mining and Crushing Gypsum at the Mine of the Blue Diamond Corporation, Ltd., Arden, Nev.: Inf. Circ. 6615. Bureau of Mines, 1932, 10 pp.

8 Bancroft, Howland, Platinum in Southeastern Nevada; Contributions to Economic Geology: U. S. Geol. Survey Bull. 430, 1909, pp. 192-199.

9 Ransome, F. L., Preliminary Account of Goldfield, Bullfrog, and other Mining Districts in Southern Nevada: U. S. Geol. Survey Bull. 303, 1907, p. 79-80.

10 Ransome, F. L., Preliminary Account of the Goldfield, Bullfrog, and Other Mining Districts in Southern Nevada: U. S. Geol. Survey Bull. 303, 1907, pp. 63-76.

11 Hill, James M., Notes on Some Mining Districts in Eastern Nevada: U. S. Geol. Survey Bull. 648, 1916, pp. 4253.

12 Hewett, D. F., Geology and Ore Deposits of the Goodsprings Quadrangle, Nevada; U. S. Geol. Survey Prof. Paper 162, 1931, 172 pp.

13 Hewett, D. F., Work cited, p. 114.

14 Hewett, D. F., Work cited.

15 Doerner, H. A., Metallurgical Possibilities of the Descloizite Ores at Goodsprings, Nev.: Rept. of Investigations 2433, Bureau of mines, 1923, 19 pp.

16 Hewett, D. F., Work cited, p. 152.

17 Hewett, D. F., Work cited, p. 146.

18 Hewett, D. F., Work cited, p. 146.

19 Koster, J., and Shelton, S. M., Electrolytic Manganese: Eng. and Min. Jour., vol. 137, October 1936, pp. 510-513.

20 Hale, Fred A.. Jr., Manganese Deposits of Clark County, Nev.: Eng. and Min. Jour., vol. 105 1918, p. 776.

21 Pardee, J. T., and Jones, E. L., Jr., Deposits of Manganese Ore in Nevada: U. S. Geol. Survey Bull. 710, 1920, pp. 209-242.

22 Hewett. D. F., and Webber, B. N., Bedded Deposits of Manganese Oxides Near Las Vegas, Nev.: Nevada University Bull., vol. 25, 1931, 17 pp.

23 Lincoln, Francis Church, Mining Districts and mineral Resources of Nevada: Mackay Sch. Mines Bull., 1923, pp. 22-23.

24 Noble, L. F., Colemanite in Clark County, Nev.: U. S. Geol. Survey Bull. 735-B, 16 pp.

25 Noble, L. F., Work cited.

26 Callaghan, Eugene, and Rubey, W. W., Mineral Resources of the Region Around Boulder Dam: U. S. Geol. Survey Bull. 871, 1936, pp. 106-113.

27 Rubey, W. W., and Callaghan, Eugene, Mineral Resources of the Region Around Boulder Dam: U. S. Geol. Survey Bull. 871, 1936, pp. 119-139.

28 Ransome F. L., Preliminary Account of the Goldfield, Bullfrog, and Other Mining Districts in Southern Nevada: U. S. Geol. Survey Bull. 303, 1907. pp. 69-72.

UNITED STATES DEPARTMENT OF THE INTERIOR
Harold L. Ickes, Secretary
GEOLOGICAL SURVEY
W. C. Mendenhall, Director

Bulletin 906-D

GEOLOGY OF THE SEARCHLIGHT DISTRICT CLARK COUNTY, NEVADA

BY
EUGENE CALLAGHAN

Contributions to economic geology, 1938-39
(Pages 135-188)

UNITED STATES
GOVERNMENT PRINTING OFFICE
WASHINGTON : 1939

HORNFELS FROM HILL 1,000 FEET SOUTH OF DUPLEX MINE.

This rock, originally a volcanic breccia, is now a very fine grained partly reconstituted aggregate, but the original fragmental character is brought out by weathering. Natural size.

VEIN MATERIAL FROM BURDICK STOPE ON LEVEL 5 OF DUPLEX MINE, BELOW NEW YEARS GIFT SHAFT.

A breccia of greenish andesite hornfels cemented by "comb" quartz stained dark by iron oxide and containing some green copper stain. Gold occurs in minute spots in the quartz.

CONTENTS

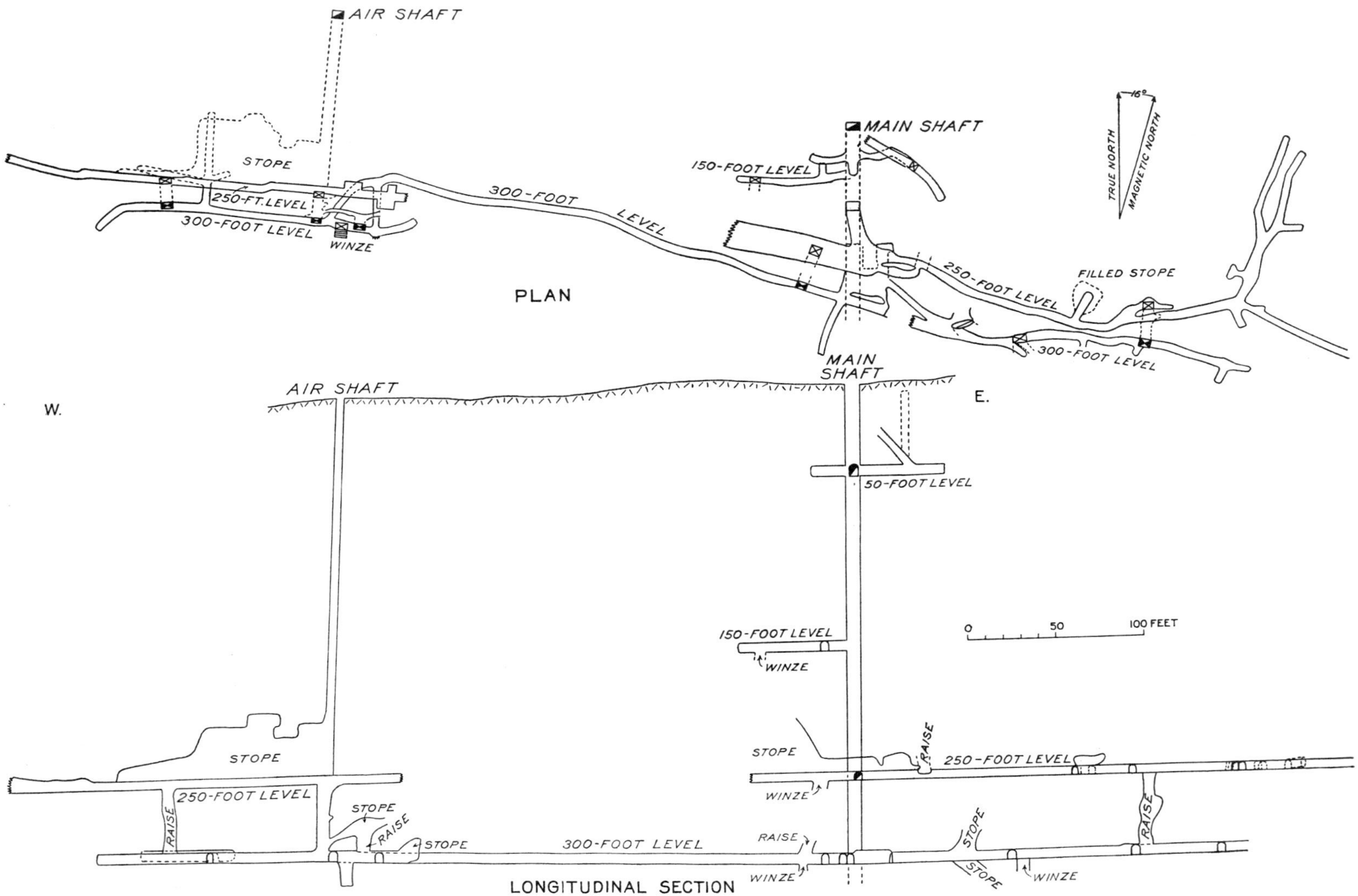

PLAN AND LONGITUDINAL SECTION OF SEARCHLIGHT M. & M. MINE.

From undated mine map.

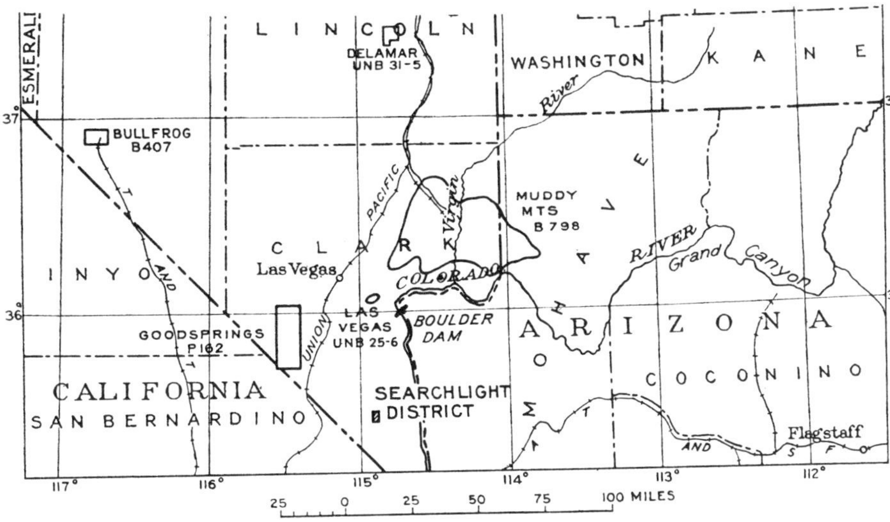

LOCATION

The Searchlight district lies on the east side of Piute Valley in the southern part of Clark County, Nev. Most of the productive mines are in secs. 22, 27, and 34, T. 28 S., R. 63 E., and secs. 2 and 3, T. 29 S., R. 63 E. A much larger area has been prospected, and several mines about 5 miles east of Searchlight are generally included in the Searchlight district, as well as some more distant to the northeast. The town of Searchlight in 1930 had a population of 137, but in 1931 the population was probably considerably larger owing to mining activity and to anticipation of the possible location of the Colorado River aqueduct through the district. A branch of the Atchison, Topeka & Santa Fe Railway once extended to Searchlight from Barnwell, but the track was taken up in 1924. The district is accessible by several roads, the most important of which extend 22 miles from Nipton on the Union Pacific Railroad, 56 miles from Needles, Calif., and 55 miles from Las Vegas, Nev. Searchlight is 38 miles in a direct line from Boulder Dam.

SURFACE FEATURES

The Searchlight district covers a group of hills that, topographically, form a western outlier of a relatively low range of mountains. These mountains trend roughly north and south and lie along the west side of the Colorado River from the vicinity of Boulder Dam to the vicinity of Needles, Calif. They are bounded on the west by Piute Valley, which drains into the Colorado River near Needles, and on the north by a dry-lake valley. The divide between the Piute Valley drainage and that to the north lies at the north side of the area shown on the geologic map (pl. 41). Pediments are well developed in many places along the base of the mountains, both on the west slope and on the slope toward the Colorado River. The term "pediment" is applied to an erosion surface, which slopes gently away from the base of a mountainous area and over most or all of which bedrock is exposed.

The altitude of the town is about 3,500 feet, and the nearby hills rise from 200 to 700 feet above the pediment surface, which merges with the gentle slopes of Piute Valley and the dry-lake valley to the north. Most of the mines are in the pediment or in the gently rolling hills that rise 20 to 200 feet above it. The boundary between pediment and alluvial fill is not everywhere readily distinguished, and no boundary is shown on the map (pl. 41). A shaft west of the Searchlight M. & M. mine reveals that there is a valley filled by more than 200 feet of alluvium even with the pediment surface. The distance between the Searchlight Parallel shaft, in the pediment on the east side of the wash, and a shaft in a pediment on the west side is only 0.6 mile. The buried valley probably lies between them. It is an open question whether the buried valley represents a surface cut prior to the formation of any pediment whatever, or whether it represents a channel in an older pediment, which was later filled with debris even with the pediment surface. Pediments are particularly prominent in this region and indicate that there has been a long period of erosion without local differential deformation.

CLIMATE AND VEGETATION

The district is included in the Mohave Desert region [5] and has the climatic features of other parts of the arid Southwest. It receives a rainfall slightly greater than that of many parts of this region, for according to Thompson [6] the average annual precipitation for 8 years ending June 30, 1922, was 9.41 inches, though the average for 6 of these years was about 7.4 inches. Water is obtained from deep wells or mine shafts, and the vertical depth to water commonly ranges from 100 to 300 feet. Information obtained by Hewett [7] shows that ample water is available for milling but not enough for irrigation. Some of the mine water, particularly that from the Santa Fe and Pompeii shafts, is potable. Owing to the altitude (3,500 feet), temperatures are not so high as, for example, along the Colorado River, and moderately cold weather is experienced in winter. The desert vegetation, consisting mainly of Joshua trees, Spanish bayonet or yucca, creosote bush, and several varieties of cactus, is fairly abundant.

GEOLOGY

GENERAL FEATURES

The most conspicuous formation in the district is a large body of quartz monzonite, which has intruded a group of older lava flows and breccias, and a body of gneiss, which extends southward from

[5] Thompson, D. G., The Mohave Desert region, California: U. S. Geol. Survey Water-Supply Paper 578, pp. 716–727, 1929.

[6] Idem, pp. 82, 718.

[7] Hewett, D. F., Ground water in Piute Valley, Nevada: U. S. Geol. Survey Press Memo., August 23, 1922; Thompson, D. G., op. cit., pp. 720–27.

the Quartette mine, in the southern part of the district. Dark-brown or gray andesite porphyry with conspicuous plagioclase phenocrysts occurs around the border of the quartz monzonite and has likewise invaded in a most intricate fashion the older rocks as dikes, sills, and irregular-shaped masses. The porphyry is in turn invaded by the quartz monzonite. A distinctly younger group of lavas and tuffs lies upon an eroded surface of the quartz monzonite and earlier rocks, and bedded arkose lies between the older and the younger volcanic rocks northeast of the Blossom mine. Only faulted remnants of the younger lavas occur on hills in the northern part of the district, but they make up the hills northwest of the district. Red volcanic rocks half a mile west of the Searchlight Parallel mine probably belong to the younger group.

The ages of the different rocks are unknown. The gneiss is generally regarded as pre-Cambrian. The quartz monzonite may be early Tertiary and closely related in age to large areas of quartz monzonite in the Ivanpah quadrangle, west of Searchlight, described by Hewett.[8] The andesite porphyry is believed by the writer to be an early border facies of the quartz monzonite. The intruded volcanic rocks may be very early Tertiary or possibly older. The later volcanic rocks are probably closely related to the late Tertiary volcanic rocks in the Ivanpah area.[8] Hewett [9] has also pointed out the possibility of three groups of volcanic rocks in this region. Certainly two groups of volcanic rocks separated by a great lapse of time are present in the Searchlight area. More extensive geologic mapping in this area will probably clarify these relations.

The veins in the district are restricted to the older group of rocks (gneiss, older volcanics, and andesite porphyry). The fact that they are distributed around the margin of the quartz monzonite body suggests a genetic relation. However, the quartz monzonite body may be involved only so far as it was a competent mass against which the other rocks were broken to form the fissures. The fissures may have been mineralized from another source at a much later time. Areas of altered rock and some silicified material occur in the later volcanic rocks, but this alteration is not believed to be related to the principal veins of the Searchlight district

Deformation occurred at the time of intrusion of the andesite porphyry and on a less extensive scale at the time of the intrusion of the quartz monzonite. Apophyses of the quartz monzonite tend to follow the direction later taken by the veins. Subsequent regional movements caused the formation of the vein fissures. Deformation subsequent to the emplacement of the veins has caused both relative

[8] Hewett, D. F., Geology and ore deposits of the Ivanpah quadrangle, Nevada and California: U. S. Geol. Survey report (in preparation).

[9] Hewett, D. F., personal communication.

movement between the walls of the veins and minor faulting which has displaced parts of the veins, particularly in the Duplex and Good Hope mines. The younger volcanic rocks were involved in the later deformation and were faulted and inclined to the northwest in the hills northwest of the district. The extensive pediments indicate that little or no local deformation has taken place in this area in recent geologic time.

ROCK TYPES

GNEISS

A body of gneiss derived from granite occurs in the hill at the south end of the Quartette vein and in the pediment to the south. The original nature of the rock is assured, for sheared pegmatites and aplites were observed within the gneiss.

The gneiss is largely brown in the outcrop but pinkish gray with discontinuous brown streaks on the fresh surface. Foliation is well developed but the rock is not broadly banded. Feldspar forms lenticular grains, mostly less than 2 millimeters in diameter, that are surrounded by quartz in mosaic aggregate. Many of the feldspar grains have granulated margins. Except for scattered grains of oligoclase, the feldspar is wholly orthoclase. There is some evidence of replacement of plagioclase by orthoclase. The dark streaks consist of extremely fine-grained magnetite, brown iron oxide, chlorite, and epidote. The chlorite is evidently an alteration product of biotite, which is preserved in some places. Apatite is abundant, and zircon and rutile were observed. A pegmatitic facies of the gneiss, illustrating the type of banding, is shown in plate 42.

OLDER VOLCANIC ROCKS

The older volcanic rocks that were invaded by both the andesite porphyry and the quartz monzonite have been considerably modified by heat and solutions given off by the invading magmas. The original rocks appear to have been chiefly andesite flows and breccias (pl. 43). No original basalts or rhyolites were definitely recognized. These volcanic rocks were probably deposited on an erosion surface, part of which was undoubtedly cut on the gneiss, which is much older. Through contact or thermal metamorphism these volcanic rocks have been changed to a dense, hard rock, chiefly by the addition of quartz and the formation of epidote or clinozoisite with minor quantities of magnetite, biotite, and chlorite. This type of metamorphism is broadly restricted to a zone along the contact of the intrusive body but varies greatly from place to place as local structural conditions affected the movement of these high-temperature solutions. The term "hornfels" is used for these partly metamorphosed volcanic rocks as they have been converted to a very fine grained aggregate in which the original texture and structure are largely obliterated and new minerals formed.[10]

These hornfelses are mostly dark greenish gray and very fine-grained. Differential weathering of exposed surfaces reveals flow structure and phenocrysts in the lavas and the fragments in the breccias (pl. 44). No definite evidence of shearing of these flows was found. Plagioclase phenocrysts, chiefly andesine, are moderately well preserved, but all original ferromagnesian minerals have been destroyed. In some places plagioclase phenocrysts are sufficiently abundant and large to cause confusion with some facies of the andesite porphyry. The groundmass is largely recrystallized and consists chiefly of remnants of feldspar, quartz, and epidote or clinozoisite. Magnetite is scattered through much of the hornfels and also occurs in bunches. The epidote or clinozoisite scattered through the rock is only slightly if at all colored in thin section, although that in later veinlets and associated with quartz and calcite is more highly colored or more ferriferous. Hornfels from a shaft south of the Blossom mine contains disseminated pyrite. In the walls of the larger veins, particularly in those of the Duplex mine, there has been little alteration of the hornfels, and vein minerals lie against recognizable feldspar grains. Weathering along the veins has caused some of the hornfels to become soft and clayey.

ANDESITE PORPHYRY

Andesite porphyry is the wall rock of many of the veins. This association has led to the belief among the miners and prospectors that ore occurs only with the porphyry or that there is a genetic relationship. It is very abundant around the margin of the quartz monzonite body, and scarcely any large fracture could have formed without being in part in the andesite porphyry. The porphyry has penetrated the gneiss and the older volcanic rocks in a most complex fashion as dikes, sills, and masses having very irregular margins. The intrusive relations give the impression of a huge breccia of older rocks with some blocks hundreds of feet long cemented by andesite porphyry. Both the andesite porphyry and the older rocks are penetrated by dikes of quartz monzonite, as shown on the map (pl. 41), but not to the extent to which the porphyry has penetrated its antecedents. It is suggested that the andesite porphyry is an early facies of the quartz monzonite intrusive bodies, and that there was only a relatively slight lapse of time between the intrusions of the two. The porphyry has been somewhat modified by the quartz monzonite intrusion and by the vein-forming solutions.

The andesite porphyry is mostly brownish on weathered surfaces, with conspicuous phenocrysts of white plagioclase, chiefly oligoclase,

[10] Buddington, A. F., and Callaghan, Eugene, Dioritic intrusive rocks and contact metamorphism in the Cascade Range of Oregon: Am. Jour. Sci., 5th ser., vol. 31, No. 186, p. 443, 1936.

as shown in plate 42. In some places, particularly in the vicinity of veins underground, it is gray, but near the Chief of the Hills mine, east of Searchlight, it has a purplish cast. There is also a wide variation in size and abundance of the plagioclase phenocrysts, which range from a fraction of a millimeter to more than an inch in length. Porphyry with only small phenocrysts is difficult to distinguish in the field from porphyritic andesite of the older volcanic rocks, on the one hand, and from very porphyritic facies of the quartz monzonite, or even from some of the porphyritic andesites in the younger volcanic rocks, on the other hand. Flakes of biotite remain as phenocrysts in much of the porphyry, as well as pseudomorphs after hornblende. The hornblende appears to have been resorbed during the cooling of the magma, rather than destroyed during later alteration. Its place is taken by a granular aggregate made up in part of magnetite and epidote, though in many specimens the form of the original hornblende is well preserved. Some of the pseudomorphs are 5 millimeters or more in length, but most are between 0.5 and 1 millimeter. The groundmass is a very fine grained aggregate composed largely of plagioclase grains.

Evidence of the initial alteration involving the resorption of hornblende was observed in all specimens. Epidote is associated with quartz in nodules in the porphyry. A good exposure at the Chief of the Hills mine shows a broad band of epidotized porphyry near the vein, indicating that at least part of the epidote-forming solutions arose along the vein fractures, apparently prior to the deposition of the vein minerals. Some of the porphyry near the veins is changed to a soft aggregate consisting largely of clay minerals. Porphyry associated with some of the veins in the northern part of the district, particularly the Pompeii, is largely changed to adularia.

It is suggested that the andesite porphyry represents the first invasion of the magma into a relatively thin cover of volcanic flows and breccias that broke and separated readily because of a relatively light superincumbent load. Probably the porphyry magma broke through to the surface and accumulated as flows to such a depth that the next pulse of the magma had sufficient cover to cool as quartz monzonite.

QUARTZ MONZONITE AND ASSOCIATED DIKE ROCKS

The quartz monzonite body underlies a large area east of the main part of the district and north of the mines east of the town. It is part of a large body or group of bodies that are known to occur in this region but have not been mapped. The contact is very irregular, as shown by plate 41. Many dikes and sill-like protuberances project into the invaded rocks. Only the margins of the body and dikes extending outward from it were studied, as the veins are either wholly outside the quartz monzonite or extend only a short distance into it.

The marginal facies of the quartz monzonite is mostly porphyritic, and both the size and the relative proportions of phenocrysts as well as the proportions of dark minerals and color are variable. The rock consists of plagioclase (oligoclase and andesine), orthoclase, quartz, biotite, augite, magnetite, apatite, and zircon. In hand specimens the orthoclase is light gray, and much of the plagioclase is dark purplish gray. Quartz is either interstitial to the feldspars or in graphic intergrowths with orthoclase. Biotite is the most prominent ferromagnesian mineral, but augite is moderately abundant. Magnetite grains are widely variable in size. A slight amount of orthoclasization of plagioclase has taken place. Some biotite has become green, and more augite is partly altered to chlorite. Some areas of chlorite give no definite clue to the original mineral, which may have been hypersthene.

Approximately equal proportions of orthoclase and plagioclase led Ransome [11] to classify the rock as a quartz monzonite, though the dioritic aspect imparted by the augite should be noted. No hornblende was seen in any of the thin sections of the marginal facies.

Dikes of granophyre, which consists largely of graphic intergrowths of quartz and orthoclase, were observed; also dikes of lamprophyre or malchite, and light-gray rhyolite. The rhyolite dikes are especially prominent in the northern part of the district. The malchite consists largely of a groundmass of minute altered feldspars and sporadic plagioclase phenocrysts and pseudomorphs of chlorite and carbonate after a former ferromagnesian mineral. Chlorite, carbonate, and epidote now make up a large portion of the rock The rhyolite consists of scattered phenocrysts of quartz, oligoclase, and chlorite pseudomorphs after biotite in an extremely fine-grained devitrified groundmass. Sericite is abundant as an alteration product. The rhyolite dikes are probably related to the later volcanic rocks. A group of biotite andesite dikes that traversed the andesite porphyry was noted by Jaggar and Palache [12] in the Quartette mine. Their age relation to the veins is not known. Locally they were called neoandesite.

LATER VOLCANIC ROCKS

A group of volcanic rocks, consisting of both fragmental or pyroclastic rocks and flows, constitutes the youngest group in the small area studied. These later volcanic rocks are exposed on hills in the northern part of the district, in the more prominent hills to the west of the northern part of the district, in the pediment west of the district, in a hill south of the Quartette mine, and in a wedge-shaped area west of the Chief of the Hills mine. In a gully northeast of the Blossom mine a small lens of arkose lies between exposures of the later volcanic rocks

[11] Ransome, F. L., Preliminary account of Goldfield, Bullfrog, and other mining districts in southern Nevada: U. S. Geol. Survey Bull. 303, p. 66, 1907.

[12] Jaggar, T. A., Jr., and Palache, Charles, private report on Quartette mine, 1911.

and the quartz monzonite and associated early, partly metamorphosed volcanic rocks.

The flows of the later volcanic rocks range from basalt to rhyolite, though flows of basalt are distinctly in the minority. A flow of olivine basalt, inclined toward the northwest, forms a prominent ridge northwest of the north end of the district and about 2 miles from the Pompeii mine. The basalt consists of phenocrysts of labradorite, olivine, augite, and magnetite in a very fine-grained groundmass of minute plagioclase laths and grains of ferromagnesian minerals.

Biotite andesite on a hill northwest of the Blossom mine is a light-gray rock with numerous large phenocrysts of plagioclase and smaller phenocrysts of biotite and hornblende. A reddish andesite with pronounced flow structure on a low ridge farther to the northwest contains phenocrysts of plagioclase and biotite, "ghosts" of hornblende crystals, and remants of a few augite crystals largely altered to carbonate in a groundmass consisting of minute plagioclase laths and extremely small grains of other minerals. Near the road west of the Blossom mine a rhyolite at the contact of the younger volcanics with the older rocks contains numerous quartz grains of various sizes, plagioclase phenocrysts, and smaller phenocrysts of biotite, magnetite, and titanite in a devitrified groundmass containing quartz grains and extremely minute feldspars. An andesite at the contact east of the road northwest of the Pompeii mine contains phenocrysts of biotite, hornblende, augite, plagioclase, and magnetite in a very fine-grained groundmass. Fragments of pitchstone were found in tuff near the flow of olivine basalt northwest of the Pompeii mine.

In general, the younger volcanic rocks are more rhyolitic and latitic than the flows intruded by the quartz monzonite, and they contain flows of olivine basalt.

STRUCTURE

The area studied is so small that little can be said of the regional structural setting. Bedding planes in the older volcanic rocks were largely destroyed during metamorphism and intrusion. Jaggar and Palache [13] state in this connection that the flows at the Quartette mine lie on the gneiss and dip 40° N. Original structure in both the gneiss and the older volcanic rocks is much disturbed by the seemingly heterogeneously oriented dikes, sills, and masses of andesite porphyry. At one place on the 400 level of the Quartette mine the foliation of the gneiss strikes N. 34° W. and dips 15° NE. The quartz monzonite forms an almost rectangular body within the district, with one corner lying under cover east of the Quartette mine and another corner well exposed in the northern part of the district. Apophyses of the quartz monzonite tend to follow the west-northwest

[13] Jaggar, T. A., Jr., and Palache, Charles, private report on Quartette mine, 1911.

trend later taken by the veins, and at the south end of level 2 from the Searchlight shaft of the Duplex mine a dike of quartz monzonite lies parallel to the vein fracture. The quartz monzonite seems to have resisted rupture, and as a consequence, so far as could be ascertained, the older rocks molded to it are not greatly faulted. During regional movements that followed the consolidation of the quartz monzonite, that body appears to have acted as a competent mass, so that fractures were formed in regular pattern around its margin in the less competent older volcanic rocks and andesite porphyry.

Most of the fractures subsequently mineralized to a varying degree have a west-northwest trend, but there is a much less prominent and less extensively mineralized system nearly at right angles. Both sets of fractures are believed to be of the gash type, with maximum movement in the center and the amount of displacement decreasing both toward the ends and in depth. This seems to account for the restriction of the ore bodies. Movements subsequent to the deposition of the ores caused relative displacement between the walls of the veins, brecciating the vein material and forming gouge seams through both the vein material and the country rock. These movements also caused minor displacements of the veins along normal faults striking slightly east of north and dipping mostly to the east.

The later volcanic rocks are greatly disturbed. Those northwest of the north end of the district are inclined to the northwest as much as 55° and appear to be extensively faulted. Ransome [14] shows a wedge of volcanic rocks inclined toward the west between the town and the mines in the eastern part of the district.

In general, it appears that the quartz monzonite and the older rocks immediately associated with it constitute an uplifted mass that has resisted deformation. Regional stresses acting against the quartz monzonite body produced the fractures around that body that were later mineralized. Later uplifts caused faulting in the mineralized area and faulting and tilting of the later volcanics away from the older rocks.

SUMMARY OF GEOLOGIC HISTORY

From such information as is available in this small area the following sequence of geologic events is deduced. The pre-Cambrian (?) gneiss was denuded of the great thickness of Paleozoic and Mesozoic rocks exposed to the west [15] and north.[16] On this floor of the pre-Cambrian (?) rocks, which was probably very uneven, there was

[14] Ransome, F. L., op. cit., p. 67.

[15] Hewett, D. F., Geology and ore deposits of the Goodsprings quadrangle, Nevada: U. S. Geol. Survey Prof. Paper 162, pp. 9-36, 1931; Geology and ore deposits of the Ivanpah quadrangle, Nevada and California: U. S. Geol. Survey report (in prepartion).

[16] Longwell, C. R., Geology of the Muddy Mountains, Nevada: U. S. Geol. Survey Bull. 798, pp. 20-68, 1928.

extruded a series of andesitic lava flows and tuffs or volcanic breccias of unknown thickness. These rocks were intruded by a magma that invaded both the gneiss and the volcanic rocks in a most intricate pattern. The first intrusions consolidated as andesite porphyry, but at a later stage a part of the magma invaded the porphyry as well as the older rocks and solidified as quartz monzonite. It is suggested that both the texture of the andesite porphyry and the pattern of the intrusive masses indicate that the magma first broke into a thin cover of volcanic rocks. It may even have reached the surface and spread out as flows, affording the later magma sufficient cover to crystallize as quartz monzonite. The early lavas were changed to hornfels by heat and solutions from the intrusive rocks. The andesite porphyry was also somewhat modified as a consequence of the intrusion of the quartz monzonite.

Regional movements acting against the competent buttress afforded by the quartz monzonite body produced the fractures that were later mineralized. The mineralization may have been closely related to the quartz monzonite in that the solutions came from very deep within that body after the magma near the surface had consolidated. However, the mineralization may have been related to a later igneous epoch, its relation to the quartz monzonite being determined only by the previously available fracture system.

The area was uplifted and deeply eroded, probably with accompanying movements on the veins and brecciation of the ores. Possibly the oxidation of the veins took place at this time, when the relief was great, providing for circulation of ground water and accounting for the lack of coincidence between oxidation and present ground-water level. Another series of lavas and fragmental rocks was laid down on the erosion surface, and dikes of rhyolite and possibly of andesite penetrated the older rocks, including the quartz monzonite. Large masses of the volcanic rocks were altered by hydrothermal solutions. Further uplift of the central area tilted the later volcanic rocks to the west on the west side of the area. Faulting, both along and across the veins, probably accompanied this later uplift but may have taken place at an earlier time. Erosion progressed, Piute Valley was partly filled, and pediments were developed.

As no fossiliferous rocks occur in this area and no other methods of dating have been tried, it is not possible to fit these events into the geologic timetable exactly. As volcanic rocks in this region are generally believed to be of Tertiary age, all the rocks later than the gneiss are probably Tertiary. Granitoid intrusive rocks in the area to the west[17] are probably of early Tertiary age and earlier than the ores, whereas the younger volcanic rocks are much later than the

[17] Hewett, D. F., Geology and ore deposits of the Goodsprings quadrangle, Nevada: U. S. Geol. Survey Prof. Paper 162, p. 38, 1931; Geology and ore deposits of the Ivanpah quadrangle, Nevada and California: U. S. Geol. Survey report (in preparation).

ores. Hewett [18] has noted two groups of volcanic rocks that he has assigned to the Miocene and Pliocene, both of which are later than the quartz monzonite intrusions and most of the metalliferous deposits. He has also pointed out the possibility of a third, much earlier group of lavas, which may be represented by the contact-metamorphosed lavas of the Searchlight district. Some of the intrusive rocks in Utah [19] are assigned to an Eocene or early Oligocene age, and some of the intrusions in the Stockton and Fairfield quadrangles [20] are intrusive into earlier volcanic rocks. In the Pioche area, Gillson [21] regards the volcanic rocks intruded by quartz monzonite as of Miocene (?) age, so that the intrusives may be mid-Tertiary. Small dioritic intrusive bodies in the Delamar district [22] invade Paleozoic rocks and are earlier than some if not all of the volcanic rocks there. Small dioritic intrusive bodies in the Chief district [23] are associated with ores in Paleozoic rocks. Ferguson [24] has classified the ore deposits of the Searchlight and Eldorado Canyon districts in the group associated with Tertiary lavas.

Such observations and correlations suggest that the quartz monzonite and older lavas are not later than mid-Tertiary and are probably early Tertiary. The mineralization may have followed closely the solidification of the quartz monzonite, but that it was as late as mid-Tertiary time is altogether possible. The later volcanic rocks correspond, particularly in their content of olivine basalt flows, with the group generally regarded as of Pliocene age.

MINERAL DEPOSITS

HISTORY AND PRODUCTION

The Searchlight district is one of the later discoveries in Nevada, having been located in 1897, over 30 years after the initial development of the Eldorado Canyon district nearby. It has probably yielded gold and silver ores every year since development began, in 1898, and has a continuous record of production since 1902. Copper and lead have also been produced every year from 1905 to 1934 except 1930, according to data in Mineral Resources of the United States.

Legend has it that the first discovery was made on the Searchlight vein by G. F. Colton, who named it after the trademark on a box of

[18] Hewett, D. F., oral communication.

[19] Gilluly, James, Geology and ore deposits of the Stockton and Fairfield quadrangles, Utah: U. S. Geol. Survey Prof. Paper 173, p. 91, 1932. Nolan, T. B., The Gold Hill mining district, Utah: U. S. Geol. Survey Prof. Paper 177, p. 48, 1935.

[20] Gilluly, James, op. cit., p. 91.

[21] Gillson, J. L., Petrography of the Pioche district, Lincoln County, Nev.: U. S. Geol. Survey Prof. Paper 158, p. 84, 1929.

[22] Callaghan, Eugene, Geology of the Delamar district, Lincoln County, Nev.: Nevada Univ. Bull., vol. 31, No. 5, pp. 24–25, 1937.

[23] Callaghan, Eugene, Geology of the Chief district, Lincoln County, Nev.: Nevada Univ. Bull., vol. 30, No. 2, pp. 32, 1936.

[24] Ferguson, H. G., The mining districts of Nevada: Econ. Geology, vol. 24, p. 135, 1929.

matches. According to an engineer's report, the area including the Duplex mine was located by Colton on May 6, 1897, as the IXL copper mine. The Duplex property is credited with the first shipment of ore from the district, though it did not become the principal producer until 1917. Exploration of the Quartette vein was started in 1898, and good ore was first found in 1899 at some distance below the surface. The Quartette Mining Co., in which Col. C. W. Hopkins, of Boston, was the principal figure, acquired the Quartette, Boston, and Chief of the Hills mines. A 20-stamp mill and cyaniding plant was under construction in 1900 on the Colorado River for the treatment of Quartette ore, but according to Ransome [25] it was not operated until 1902. The mine was connected with the mill by 15 miles of narrow-gage railroad track. A strike in 1903 lasting about 3 months suspended operations throughout the district. Shortly thereafter water was struck in the Quartette mine. The mill on the river was then moved to the mine and subsequently enlarged to 40 stamps. There were 600 people in the town in 1903,[26] and three mines, the Quartette, Duplex, and Southern Nevada, were equipped with stamp mills. Most of the veins were discovered and partly explored during the first few years. Some of the mines, such as the Blossom, had largely been worked out by 1906. The boom period lasted until 1910. According to the census, the population of the Searchlight precinct in 1900 was 211, in 1910 it was 613, of whom 387 were in the town, in 1920 it was 161, and in 1930 it was 137.

The Quartette mine produced from $200,000 to $400,000 a year over a period from 1903 to 1909. In 1910 the production began to decline, and in 1911, after considerable exploration on the lower levels, company operation ceased and a leasing system was established. In 1913 the 40-stamp mill was burned, and the mine was sold to the Searchlight Mercantile Co. The leasing system was maintained, and the Quartette continued its leadership in the district until 1917. Steady production was maintained until 1921, and occasional leases have produced a small amount of ore since that time. In 1934 the Mollin Investment Co. constructed a flotation mill to handle old tailings and did some exploratory work on the 400-foot level. The Duplex mine was operated occasionally by companies, but most of the ore was taken out by lessees between 1915 and 1930. In 1930 the Searchlight Gold Corporation took over the property, erected a 100-ton flotation mill, and worked tailings, material from the dump, and pillars and small stopes from the mine. A considerable production was obtained, particularly of lead. Other mines have been operated intermittently, but most of the production was obtained in the first

[25] Ransome, F. L., op. cit., p. 64.

[26] Searchlight district: Eng. and Min. Jour., vol. 75, p. 347, 1903.

years of operation. A branch of the Atchison, Topeka & Santa Fe Railway was extended to the district in 1907, but it was discontinued in 1924.

The recorded production, taken from the annual chapters on Nevada by V. C. Heikes and C. N. Gerry in Mineral Resources of the United States, is given in the accompanying table. The total value of the metals produced (gold, silver, copper, and lead) from 1902 to 1934, inclusive, is $4,657,711. Of this amount $3,239,588, or over 70 percent, was produced in the years 1902–10. The Quartette mine has accounted for the greater part of the production of the district, particularly of gold and copper. The Duplex mine, especially in later years, has been the principal lead producer. The production of the district prior to 1902 is unknown but probably did not exceed $500,000, and some later production has undoubtedly not been recorded. The number of producers of small amounts of gold and silver ore has greatly increased since the increase in the price of gold.

Metal production of Searchlight district, Nevada

[Data from Mineral Resources of the United States and Statistical appendix to Minerals Yearbook]

Year	Mines reporting	Ore (tons)	Gold (ounces)	Silver (ounces)	Copper (pounds)	Lead (pounds)	Total value
1902	3	10,910	6,146.52	1,175	----------	----------	$127,676
1903	3	19,522	19,274.97	27,691	----------	----------	410,079
1904	3	16,750	18,400.17	13,498	----------	----------	388,068
1905	6	38,069	19,329.42	28,528	22,808	12,064	420,931
1906	5	45,668	25,144.60	11,543	11,182	9,655	530,227
1907	7	45,921	23,440.98	7,494	37,063	38,113	498,947
1908	7	52,193	13,137.34	10,883	14,954	44,857	281,199
1909	8	68,931	16,236.83	13,544	22,916	36,209	347,224
1910	13	27,331	10,406.24	11,489	93,848	45,847	235,237
1911	10	1,850	966.49	2,136	12,095	7,659	22,968
1912	9	4,158	2,237.04	2,562	10,126	10,032	49,942
1913	15	5,989	6,653.82	5,903	53,971	61,006	152,161
1914	5	3,057	3,998.56	2,855	5,556	21,919	85,830
1915	13	7,766	4,984.68	3,546	20,970	35,562	110,182
1916	15	6,322	3,902.74	4,340	24,392	9,951	90,220
1917	14	12,866	3,444.94	9,073	98,923	59,453	110,808
1918	14	1,700	2,649.07	8,395	44,477	77,863	79,670
1919	9	6,980	2,8[illegible]4.37	5,502	18,074	39,307	70,612
1920	7	2,131	3,432.36	5,275	44,941	91,436	92,287
1921	9	1,182	3,600.64	9,877	31,128	98,048	92,737
1922	7	1,441	1,661.28	5,024	12,847	44,375	43,540
1923	6	1,142	1,006.96	2,708	22,150	47,651	29,628
1924	2	850	830.14	1,267	5,801	46,050	22,453
1925	5	2,833	3,313.92	2,586	14,974	86,768	79,975
1926	4	1,881	1,879.41	1,720	7,430	81,977	47,522
1927	4	106	334.90	306	1,429	10,415	7,940
1928	3	77	223.78	252	2,968	9,464	5,573
1929	3	184	105.28	58	654	2,743	2,495
1930	3	124	19.42	27	----------	----------	411
1931	8	20,855	3,276.30	5,826	6,022	274,687	80,129
1932	7	2,973	2,018.41	4,853	4,353	355,458	54,031
1933	10	2,996	912.58	1,182	2,261	12,040	19,869
1934	25	6,701	1,745.52	8,478	2,237	4,951	67,140
Total	----------	421,459	207,569.78	219,596	650,550	1,675,560	4,657,711

MINERALOGY AND CLASSIFICATION

The principal product of value in the veins of the Searchlight district has been gold, though some silver has been recovered from all of the mines. Four of the mines—the Quartette, Duplex, Good Hope, and

Big Casino—have produced considerable lead and copper. In terms of recovered metals, 6 out of 10 mines yielded a larger proportion of gold than silver, mostly 2 to 1 by weight. The Duplex mine yielded more gold than silver by weight during the first few years of production, but the proportions were reversed during most of the later years. The product of the Quartette mine was highly variable, but mostly contained three times as much gold as silver by weight. The Cyrus Noble, Big Casino, and Searchlight Parallel mines produced from two to nine times as much silver as gold by weight during most of their years of operation. The amount of lead in the product of the Quartette mine was about two-thirds that of copper, but the Duplex mine produced over three times as much lead as copper. Lead was more abundant than copper in the product of the Good Hope and Big Casino mines. To what extent these variations were due to differences in the processes of mining and milling used is not known.

With the exception of an unknown proportion of the ore of the Big Casino mine, the entire production of the district has come from oxidized or weathered ores. Weathering and leaching of vein materials has extended to the greatest depths attained in mining in all the veins of the main part of the district (pl. 41). Weathered vein material was found in the Quartette mine in the lowest levels, more than 800 feet vertically below the present surface and more than 600 feet below water level when the mine was first developed. No zone of secondary sulphides is known to have been found in any of the mines. Galena is the only primary sulphide remaining to any extent in the weathered veins, where it occurs as scattered lumps surrounded by its weathering products from a position within a few feet of the surface to the greatest depths of the ore shoots. Jaggar and Palache in a private report on the Quartette mine report traces of chalcopyrite and chalcocite in that mine.

Unweathered vein material probably representative of that in many of the weathered veins was found on the dump at the Big Casino mine. It consists of a breccia of country rock cemented with quartz. Most of the quartz shows comb structure, and vugs are numerous. Sphalerite, galena, and chalcopyrite occur in the quartz, the first two in about equal proportions and the chalcopyrite mostly subordinate. Some of the country rock is silicified, contains scattered areas of sericite, chlorite, and carbonate, and has all three sulphides disseminated through it. Chalcopyrite is the most abundant sulphide in the sections of silicified material that were examined. Polished surfaces of the ores, examined by C. F. Park, Jr., reveal veinlets of galena and chalcopyrite in sphalerite and isolated rounded spots of sphalerite in chalcopyrite and galena. There appears to have been an overlapping in the sequence of deposition of the sulphides, though sphalerite is the earliest, and galena appears to be slightly later than chalcopyrite. No pyrite was seen in the material from the Big Casino vein. No assays of this material are available, but an output of 4,800 tons in 1 year yielded in terms of recovered metals 0.012 ounce of gold, 0.047 ounce of silver, and 0.367 pound of copper per pound of lead, according to data compiled by V. C. Heikes, of the United States Bureau of Mines. It is not known what proportion of the material shipped was primary sulphide ore.

The oxidation products of the veins in the southern part of the district, particularly the Quartette vein, and those on the Good Hope and Duplex properties indicate that the primary ore contained appreciable quantities of sulphides. Several minor veins near these properties contain copper stain; a little copper-stained vein material occurs in veins near the Southern Nevada mine; and a very little copper stain was seen in a few pieces of vein material from the Blossom mine. The present composition of the other veins shown on the map does not indicate that they contained appreciable quantities of sulphides. The original presence of chalcopyrite is indicated by a group of copper silicates, carbonates, and sulphates. The presence of galena is indicated by cerusite, as well as by remnants of unweathered galena. The presence of sphalerite is indicated by the silicate hemimorphite (calamine) and by a minor amount of zinc, mostly less than 2 percent, in smelter returns on the shipments of ore. No definite indication of pyrite was seen, though iron oxides occur in the weathered vein material. The quartz in the Quartette vein contains abundant original specular hematite, which is not found in appreciable quantities in the other veins. Veins in the northern part of the district reveal coarse-grained calcite, as well as lamellar calcite and quartz, and probably never contained appreciable amounts of sulphides.

A wide variety of oxidation products might be expected from the weathering of complex sulphide ores in a desert region. Jaggar and Palache [27] list the following minerals from the Quartette mine: Gold, copper (traces only), galena, chalcocite, chalcopyrite (minute amounts), quartz, chalcedony, cuprite, hematite, cerusite, malachite, calcite, brochantite (basic sulphate of copper), linarite (basic sulphate of lead and copper), leadhillite (sulphato-carbonate of lead), chrysocolla, wulfenite (lead molybdate), vanadinite (lead chlorovanadate), and mottramite (cuprodescloizite—vanadate of lead, copper, and zinc). The writer also noted under the microscope the zinc silicate hemimorphite in radial aggregates in vein material from the dump at the Quartette mine. Greenish-yellow films of mottramite on other minerals are fairly common, and aggregates of vanadinite crystals are not uncommon. Orange-yellow tabular crystals of wulfenite were

[27] Jaggar, T. A., Jr., and Palache, Charles, private report on Quartette mine, 1911.

recorded as fairly abundant in the Quartette mine and were noted in the Duplex mine. The secondary copper minerals, notably chrysocolla, yield bright-green and blue colors and are outstanding. They are mostly associated with the hydrous iron oxides, and their distribution indicates that the copper has migrated widely. They occur in masses, fill fractures, and stain or replace both country rock included in the vein and quartz. However, the lead carbonate ores, which occurred in lumps and masses, some of considerable size, contained very little of the copper minerals. The average copper content of shipments of ore from the Duplex mine containing over 20 percent of lead was 1.42 percent, whereas that in ores having a lower lead content was over twice as much, or 2.90 percent. The fact that the lead carbonate ores tended to occur in lumps and masses indicates that lead did not migrate as readily as the copper under oxidizing conditions. All the calcite seen in these mines occurs as crystals and films on the secondary minerals or on surfaces of fractures. Some secondary quartz was observed as clear crystals on the surfaces of secondary copper minerals. Clay minerals, apparently mostly of the beidellite type, occur in cavities in the weathered ore and particularly in the country rocks.

Gold and silver are not ordinarily visible in any of the ores, though fine specimens of wire gold on the surface of green vein material were said to have been obtained from the Quartette mine. Under the microscope a specimen of high-grade ore from a pillar in the Burdick stope on level 5 of the Duplex mine revealed irregular grains and flakes of gold in quartz and in spots of brown iron oxides. Assays indicate that gold is more abundant in the coatings of lead carbonate adjacent to galena than in the galena itself.

The veins of the Searchlight district can be classified in various ways. All the minable ores have been of value chiefly for the gold and would be broadly classed as siliceous gold ores. All the vein material is leached and weathered, with the exception of part of that in the Big Casino mine, so that the ores could be classed as oxidized. The ores of the Quartette, Duplex, Good Hope, and Big Casino mines give evidence of an original content of the sulphides of lead, copper, and zinc, so that they could be classified as complex sulphide or base-metal veins with gold. Probably the other veins that have little evidence of original sulphides had a sparse distribution of sulphides and belong to this group. As they occur in partly metamorphosed lavas of probable Tertiary age, they are included by Ferguson [28] in the group associated with Tertiary lavas. They are associated with and are later than a large body of quartz monzonite, thus differing from most of the Tertiary mineral deposits in western Nevada. How-

[28] Ferguson, H. G., The mining districts of Nevada: Econ. Geology, vol. 24, p. 135, 1929.

ever, it must be held in mind that they may not be genetically related to the quartz monzonite, and that their position is determined by regional fracturing in less competent rocks around the border of the very competent quartz monzonite body. They also differ from many of the areas of Tertiary mineralization in lavas in western Nevada in the high proportion of base metals in the shipments and in the high proportion of gold to silver (about 1 to 1, as compared to 1 to 10 or even 1 to 100 for many other districts). Even the ore from the nearby Eldorado Canyon district is materially different, both in the gold-silver ratio and in the abundance of base metals. In general, the Searchlight ores resemble both in structure and in metal content the ores of the Cascade Range in Oregon.[29] The district has been included by Burbank [30] in the group of epithermal base-metal deposits, of which it is the only representative in Nevada.

VEIN STRUCTURE

Of 34 veins in the district, 25 strike between S. 80° W. and N. 57° W., though N. 65° W. is the most common trend. Of these 25, 3 dip north and the remainder dip south. Four veins strike about N. 40° W. and dip southwest. One vein strikes about N. 40° E. and dips southeast. Four veins trending from N. 8° W. to N. 10° E. dip west. The fault that offsets the vein in the Duplex mine strikes about N. 10° E. and dips about 25° E. The fault in the Good Hope mine strikes about N. 35° E. and is nearly vertical. A probable fault at the Blossom mine strikes N. 13° E. and dips 36°–43° E. The dominant trend is west-northwest and is followed by the veins that produce almost all the ore. Two veins that strike nearly at right angles to the general trend have had a relatively small production. The dominant dip is to the south. The angle ranges from 20° to 80°, though 45° is a common angle, and almost all the mine shafts are inclined.

The dimensions of the veins vary widely. The Quartette vein and the Rambler vein, which extends on to the southeast, have an aggregate length of nearly 3,500 feet and have been explored underground for nearly 2,100 feet. The New Years Gift vein of the Duplex mine has been explored for 1,180 feet underground. The productive part of the Quartette vein was about 900 feet long, according to old maps, and the productive part of the New Years Gift vein was about 1,100 feet long. The Quartette vein was productive to the 1,100-foot level, about 920 feet down the dip of the vein, or 740 feet vertically below the surface. Stopes extend almost to the 700-foot level of the Duplex mine, 540 feet below the surface on the incline, or 400 feet vertically

[29] Callaghan, Eugene, and Buddington, A. F., Metalliferous mineral deposits of the Cascade Range in Oregon: U. S. Geol. Survey Bull. 893, pp. 24–29, 33–35, 1938.

[30] Burbank, W. S., Epithermal base-metal deposits: Ore deposits of the Western States (Lindgren volume), p. 651, Am. Inst. Min. Met. Eng., 1933.

below the collar of the Fraction shaft. Many of the other veins were stoped only near the surface and even the barren quartz veinlets largely faded out within 200 feet vertically below the surface. Stope maps indicate that the ore shoot or group of ore shoots in the Quartette mine pitched to the east, and those in the Duplex mine pitched very gently, about 10°, to the southeast. The principal ore shoot at the Blossom mine was saucer-shaped and nearly flat. The group of ore shoots on the southward-dipping vein at the Blossom mine, though very irregular, pitched to the west.

The structural features of the veins reveal the nature of the premineral movements that opened the fissures and produced the breccias where the vein material was deposited, as well as the postmineral movements that have made new fractures extending through the ore or along footwall and hanging wall producing in many places a breccia of the ore. Thus at the present time many veins are made up of lenses of vein material and lenses or horses of country rock that are separated by gouge seams; in places the richest ore is found in such lenses. There are also quartz veins and veinlets disclosed by crosscuts within both the hanging wall and the footwall of the main part of the vein; these indicate that the primary fracture was a central large break, parallel to which were numerous minor breaks of varying magnitude and continuity. In places ore has been found on these parallel fissures. Sections from old maps indicate that large stopes were made along the top of a lens within the vein. In most of the mines the best ore was limited to a central zone a foot or two in width, but in the larger mines stopes are 10 feet or more wide, and some in the Quartette mine were 50 feet wide. In the Duplex mine and particularly in the Blossom mine, the most productive parts of the vein were those that were most nearly horizontal. The origin of these nearly flat veins is not readily understood, as they do not appear to have been controlled initially by preexisting structure, nor is there any good evidence of thrust faulting. It is evident, particularly at the Blossom mine, that only a very slight displacement has been sufficient to cause the observed brecciation of the country rock.

Most of the veins show a breccia of country rock cemented with porous and vuggy comb quartz (pl. 44). Clay minerals partly fill some of the cavities, and some of the vugs are partly lined with calcite. Secondary lead and copper minerals and iron oxides partly or wholly fill fractures and cavities in some of the veins. Postmineral movements along the veins crushed the ore, as well as the nearby country rock, particularly in the Quartette mine, so that the ore as mined was soft and tended to cave and slump in the workings. The quartz characteristically does not form large solid masses. Veins in the northern part of the district, particularly the Pompeii, contained lenses of calcite, a part of which is interlayered with quartz. Brecciation in these veins does not appear to have been extensive, but numerous quartz veinlets occur in partly or wholly adularized and silicified country rock.

ROCK ALTERATION

The country rock of the veins has been modified to a variable extent, first, by emanations that escaped during or soon after the crystallization of the quartz monzonite; second, by the vein-forming solutions, which were later and may have had a source unrelated to the quartz monzonite; and, third, by the still later supergene (descending) solutions charged with acids that were formed by weathering of the primary ores. The first type of alteration is treated in the description of the older volcanic rocks on pages 142–143. It is related geographically to the periphery of the quartz monzonite, and the modifications it imposed on the intruded rocks are similar throughout the district. The principal minerals formed—epidote or clinozoite, quartz, and magnetite—are related in their distribution to the contact of the monzonite and bear no relation to the veins, unless the fractures later occupied by the veins happened to be in existence at this early time. On the other hand, the effects of alteration by the vein-forming solutions differ from south to north and suggest a poorly defined zonal arrangement in this direction, not related to the outcrop of the quartz monzonite.

In the southern part of the district the vein-forming solutions do not appear to have appreciably modified the adjacent hornfels, andesite porphyry, and quartz monzonite. Feldspar phenocrysts in rock adjacent to veins are fairly well preserved. Sericite, chlorite, and fine-grained quartz in primary ore from the Big Casino mine indicate some degree of attack of wall rock by vein-forming solutions. Whether any of the epidote in any part of the district is related to the vein-forming solutions is not definitely known. Epidote associated with quartz and calcite in minute veinlets may possibly be related to the vein-forming period, but no definite evidence for such a relation was found. In general, the wall rocks in the southern part of the district have been relatively unaffected by the vein-forming solutions.

In the northern part of the district, particularly at the Pompeii mine, the volcanic rocks and andesite porphyry, which are the wall rocks of quartz-calcite veins, have been altered to a yellowish rock composed largely of a microscopic aggregate of adularia and quartz with remnants of primary minerals. Ferromagnesian minerals and any previously formed epidote have been eliminated from this material.

Perhaps some clue to the nature of the vein-forming solutions may be obtained from the type of wall-rock alteration, as well as from the vein minerals themselves. The lack of appreciable wall-rock alteration, particularly of the feldspars, in the southern part of the district suggests that the solutions were only weakly alkaline. The abundant adularia associated with quartz-calcite veins in the northern part of the district suggests that these solutions were strongly alkaline or at least contained considerable quantities of potash. The general lack of pyrite throughout the district suggests that all the solutions were unusually deficient in iron. However, the hematite in the Quartette vein indicates that at one stage, probably the earliest, the solutions in this particular vein contained considerable iron. This large vein probably provided more ready access for vein-forming solutions, so that the temperatures were higher than in the other veins. The temperatures may have been sufficiently high to favor the precipitation of iron as oxide, even in the presence of sulphur. However, absence of sulphides in this stage may simply imply the absence of sulphur without reference to temperature. The presence of hematite suggests furthermore that the solutions were not highly alkaline and may have been distinctly acid at the time the hematite was precipitated. Succeeding solutions that deposited metallic sulphide were probably slightly alkaline and slightly cooler.

The later volcanic rocks have been affected by alteration of a different type, in which large areas of rock were changed to clayey aggregates with a little fine-grained quartz or chalcedony. These rocks are variably colored from white to red and are commonly iron-stained. West of the Chief of the Hills mine the andesite porphyry is also altered to a soft rusty rock along a zone normal to the vein. This alteration was probably much later and had nothing to do with the veins.

The later supergene alteration near those veins that originally contained appreciable quantities of sulphides has changed much of the original feldspar that survived the attack of vein-forming solutions into aggregates of clay minerals, chiefly of the beidellite type. Such rock commonly contains numerous fractures and slickensides, and though hard when first encountered readily disintegrates when exposed to damp air. The previously adularized and silicified rock in the northern part of the district does not appear to have been greatly affected by supergene solutions.

AREAL ZONES OF THE MINERAL DEPOSITS

There is a suggestion that some features of the veins as well as wall rocks are zonally distributed in the district. This zonal distribution is illustrated chiefly in the differences between the minerals and wall-rock alteration exhibited in the Quartette mine, on the south, and the Pompeii and J. E. T. mines, on the north. The Quartette vein evidently originally contained the sulphides of lead, zinc, and copper as well as abundant specular hematite distributed through the quartz.

Wall-rock alteration was comparatively slight. The product of the mine contained more gold than silver and more copper than lead. In the product of the Duplex mine, to the north, silver is slightly dominant over gold, and lead is dominant over copper. A few of the veins as far north as the Southern Nevada contain a little copper stain that indicates an original sulphide content, but they appear to grade by diminution of original sulphides so that those farthest north contained very little. A little coarse-grained and lamellar calcite was found in the Searchlight M. & M. and Blossom veins, and the lower workings on the Pompeii vein reveal lenses of coarse and lamellar calcite with quartz. Two large calcite veins crop out on the J. E. T. property. Adularization and silicification of wall rock is evident at the Pompeii, and wall rocks in many of these northern veins are iron-stained and yellowish rather than greenish or gray, as in the southern part of the area.

It might be supposed that the two extremes represent two types and two ages of mineralization, but there appears to be a gradual transition from one to the other. Furthermore, the veins in the north still preserve the high ratio of gold to silver and thus differ materially from the calcite-bearing veins of the Eldorado Canyon district, which show a high ratio of silver to gold.

It is obvious that this suggested zoning is decidedly eccentric with respect to the quartz monzonite. It indicates that uniform heat from the quartz monzonite body did not control the apparent zoning. Either one part of the body had cooled more quickly than the other, or there is no close genetic connection between the mineralization and the quartz monzonite body in its present form. If related in any way to the quartz monzonite, the solutions may have risen from some deep-seated part of the magma long after the part now in view had solidified and cooled. The relationship may be wholly fortuitous, only the belt of fracturing having been determined by the competent quartz monzonite body, and much later mineralization having been related to some igneous mass not now exposed. Study of a much larger area would probably clarify these relations.

OUTLOOK FOR THE DISTRICT

The veins of the Searchlight district are unusual among those associated with lavas in Nevada, in yielding a relatively large production of base metals and a high ratio by weight of gold to silver, essentially 1 to 1. Like many other districts, however, after an initial period of high yield of about 10 years there was a long decline but continuous annual production to the present time. It seems that a small production, chiefly from leasing operations, may be expected for several years, particularly with the maintenance of the increased price of gold. Higher prices of silver or base metals will have little effect, as the known base-metal ore bodies have been worked out, and silver forms but a small part of the value of the metal won.

The cost of mining and milling at the Quartette mine, according to Ransome[31] was between $5 and $6 a ton in 1906, before the advent of the railroad. At that time about 75 tons of ore was hoisted daily and milled in the 10-stamp mill. No other data on costs of mining and milling or of shipments to smelters were available. Lessees at the Duplex mine are said to have tried to maintain an average gross value of about $90 a ton on shipping ore, in order to make a good profit. In the last few years the mills at the Duplex and Quartette mines have worked tailings mostly.

Apparently water has been ample to meet the needs of such mills as have operated in the district. Water for milling at the Quartette mine is obtained from the mine and from the Drake shaft, northwest of the mine. The Santa Fe shaft evidently yielded ample water for the Duplex mill from a vertical depth of about 145 feet. The Searchlight M. & M. shaft yields a large flow from a depth of about 280 feet. The Spokane shaft of the Southern Nevada mine yielded ample water for a 10-stamp mill from a depth of about 200 feet. The Pompeii shaft yields abundant potable water, estimated at 65,000 gallons a day, from a vertical depth of about 240 feet.

Explorations below the known ore shoots have been discouraging. Available data on the Quartette mine indicate that two levels were run below the bottom of the ore shoot near the 1,100-foot level without revealing further ore, though oxidation products, gouge seams, and quartz stringers continued. Long crosscuts were driven in both hanging wall and footwall without revealing more than a few barren quartz stringers parallel to the principal vein. Explorations on levels 7 and 8 of the Duplex mine revealed only quartz stringers and gouge seams, which are reported to have assayed no more than a trace of gold. Only quartz stringers and gouge seams appear on level 4 at the Searchlight shaft below the prominent ore shoot on the upper levels. Level 3 at the Blossom mine reveals minable ore shoots passing into thin quartz stringers in country rock. These narrow seams must have been the channels of access of the ore solutions, but their diminution with depth does not favor the hypothesis that they may lead to deeper ore shoots. The shafts at the Cyrus Noble and Searchlight Parallel mines also show this tendency for the grade and amount of the ore to decline with depth. Several experiences such as these throughout the district appear to indicate that the veins have been "bottomed" and that deeper ore shoots do not occur. However, no convincing geologic reason for such shallow depths for these veins can be given, and deep drilling on so large a vein as the

[31] Ransome, F. L., op. cit., p. 71.

Quartette might reveal a much deeper ore shoot. It is probably true that the faults occupied by the veins are not large. Apparently the amount of relative displacement was small, and it decreased toward the ends of the fractures as well as in depth.

The immediate hope for continuity in production from the district appears to lie in further exploration along the strike of some of the veins, exploration of faulted segments, particularly in the direction of pitch of ore shoots, and searches for ore bodies within the footwalls or hanging walls of previously worked-out veins. Further and deeper exploration on the downthrown side of faulted segments, particularly in the Duplex mine, may disclose more ore shoots. Detailed studies of individual veins, especially of branching seams, may lead to the discovery of minor ore shoots within the footwalls and hanging walls. Such explorations will probably result in a continued small production from the district.

MINES AND PROSPECTS

PRINCIPAL MINES IN THE MAIN PART OF THE DISTRICT

QUARTETTE

The Quartette mine is the southernmost in the district and has been by far the largest producer. It has accounted for 64 percent of the gold, 21 percent of the silver, 58 percent of the copper, and 13 percent of the lead recorded for the district. The total yield is probably over $2,800,000 and was produced in the years 1902 to 1921 and in 1923. The main points in the history of the mine are given on pages 150–151. The property consists of 13 patented claims and fractions, which lie mostly in secs. 2 and 3, T. 29 S., R. 63 E. The vein crops out on the Golden Treasure, Copper King, and Rambler claims.

The extensive workings of the mine shown on the accompanying map (pl. 45) were wholly inaccessible in 1931, but in 1934 shaft 3 had been cleaned out to water level (between the 600- and 700-foot levels) for the installation of pumping equipment, and new work on the 400-foot level was accessible. A short drift on the 100-foot level at the north end of the mine was also accessible. According to the report in Mineral Resources of the United States for 1911, when operation by the company ceased, the depth of the main shaft on the incline was 1,167 feet, and the total length of workings was 5½ miles. The map indicates that 13 levels explore the vein for a maximum distance of 2,100 feet. As shown by the map, the length of the part of the vein wholly or partly stoped is nearly 1,000 feet. Most of the stoping was done in the part of the vein between the main shaft and shaft 3, which are 500 feet apart at the surface but diverge in depth. Most of the stopes lay between the 100- and 1,100-foot levels, though lessees have opened stopes at the surface between the main shaft and shaft 3. The map shows that the main ore body or group of ore bodies pitched to the southeast.

The country rocks of the Quartette or Golden Treasure vein are chiefly gneiss, hornfels, and andesite porphyry. Jaggar and Palache in a private report list dikes of biotite andesite, which traverse the andesite porphyry and older rocks and which were referred to as "neoandesite." Jaggar's map, reproduced by Ransome,[32] shows the irregular contact of the gneiss and hornfels at right angles to the vein near the approximate location of shaft 3, though the workings show that the gneiss is more abundant in the footwall and the hornfels more abundant in the hanging wall. Jaggar and Palache state that there has been displacement of the contact along the vein, with the relative positions of the gneiss and hornfels as shown by the mine map (pl. 45). They also state that flows forming the hornfels lie on the gneiss and dip 40° N. The andesite porphyry cuts both the gneiss and the hornfels, so it occurs in both walls of the vein as both steeply dipping and gently dipping dikes and irregular-shaped masses. Specimens from the vein show very little alteration of country rock by the primary vein solutions, but wall rocks along much of the vein have been transformed to clayey aggregates, presumably by supergene solutions accompanying the weathering of the vein.

The vein, as revealed by workings and prospect pits on the surface, has a length of 3,300 feet. The average trend is N. 70° W., but the trend at the mine, which is on the north end of the vein, is N. 65° W. The north end curves more to the northwest. The vein is variable in trend and dip, as indicated by the plan (pl. 45) and particularly by the cross sections (pl. 46). The dip at the north end of the vein on the 100-foot level is 40° S., but it steepens to 50°–60° throughout most of the mine. Dips of 72° S. were observed at the Rambler workings on the ridge. According to the map (pl. 45), the vein was explored to the 1,300-foot level, a distance of 1,160 feet on the incline, or 868 feet vertically below the surface. Such information as is available indicates that the slip planes or postmineral fractures and some quartz, but no ore, were exposed below the 1,100-foot level.

The structure of the vein is complex, as indicated by the map and particularly by the sections. As shown in the new work on the 400-foot level (pl. 45), numerous intersecting gouge seams enclose lenses of quartzose vein matter and lenses of country rock as well. Some of the lenses of ore were stoped to widths of 50 feet, according to Ransome.[33] Crosscuts show that gouge seams and quartz veins or veinlets, mostly parallel to the main vein, occur in both the footwall and hanging wall. The sections (pl. 46) suggest that the large stopes were opened at the tops of lenses formed by large splits in the vein.

---

[32] Ransome, F. L., Preliminary account of Goldfield, Bullfrog, and other mining districts in southern Nevada: U. S. Geol. Survey Bull. 303, p. 70, 1907.

[33] Ransome, F. L., op. cit., pp. 70–71.

Ransome [34] describes the ore as follows:

The ore is ordinarily a soft shattered mass of altered country rock intimately mingled with the various ore minerals. It is generally oxidized, but a little chalcocite is said to be occasionally found, and some residual kernels of galena were seen in the west drift of the eighth level. Good ore is nearly always associated with chrysocolla or with earthy cuprite and in some places carries from 10 to 12 percent copper. Small cupriferous bunches are sorted and shipped. Where galena or, more commonly, cerusite appears the ore carries a larger proportion of silver than where copper is the dominant base metal. Wulfenite, the molybdate of lead, in characteristic square, tabular orange-colored crystals, is very common throughout the mine and does not necessarily indicate ore. It is apparently one of the later products of oxidation and occurs in little vugs and open fissures, implanted on the other minerals. Hematite is abundant but, like the wulfenite, does not seem to be particularly characteristic of good ore. Quartz is widely distributed through the lode but never forms large or solid masses. Its most typical occurrence is in small vugs and veinlets.

Jaggar and Palache in a private report state that the usual ore consisted of gold-bearing quartz more or less deeply stained with chrysocolla and in places containing large masses of lead carbonate. Their mineral list is given on page 153. Material collected by the writer from a dump near shaft 3 consisted of a porous aggregate of quartz stained mostly brown and containing chrysocolla and malachite, as well as radial aggregates of hemimorphite (calamine, hydrous silicate of zinc), visible under the microscope. Specular hematite occurs in plates and sheafs in much of the quartz but is turned red in most of the material examined. The hydrous oxides of iron are present in most of the specimens.

According to available information postmineral faults offset the vein at various places, and the vein appears to be offset on the surface east of shaft 3. Postmineral movements have brecciated the primary ore and have facilitated the weathering of the vein down as far as operations extended and well below the present water level, which in December 1934 was between the 600 -and 700-foot levels, or about 400 feet vertically below the surface. Ransome states that water was struck originally 180 feet below the surface. According to Hewett,[35] the reported water yield of the mine was 25,000 to 30,000 gallons a day.

The available information and maps indicate that a group of ore bodies extending chiefly from the 200-foot to the 1,100-foot level at the north end of the Quartette or Golden Treasure vein and attaining widths as great as 50 feet were stoped. The value of the ore varied widely, and Ransome states that ore worth \$25 to \$40 a ton, at the old price of gold, was considered good ore, though some worth \$400 a ton was being stoped in 1906. The cost of mining and milling

[34] Ransome, F. L., op. cit., p. 71.

[35] Hewett, D. F., in Thompson, D. G., The Mohave Desert region, California: U. S. Geol. Survey Water-Supply Paper 578, p. 720, 1929.

PLAN AND STOPE MAP O

Geology in the new (1934) work on the 400-foot level and at the north end of the 100-foot level by Eugene Callaghan, 1934. Base

UARTETTE MINE.

print bearing date October 13, 1910. Notes on 1,200- and 1,300-foot levels from old prints loaned by R. T. Walker.

was between $5 and $6 a ton before the completion of the railroad. Mining was carried on by the company until 1911, when extensive explorations failed to yield new ore bodies. The mine was operated under a leasing system from 1911 to 1921, and lessees since that time have produced a little ore from rock near the surface. The ground is reported to have been soft and required extensive timbering. Almost all the levels are closed, and any new exploration in a mine that was probably extensively explored over so long a period of productive history would be very costly, with little assurance of revealing sufficiently large bodies to mine. Nothing is known to the writer of any successful development of ore shoots in the eastern or Rambler portion of the vein, nor is it known how extensive explorations in that area may have been. The vein is so long, however, that the possibility of an ore shoot in this area should be entertained.

GOOD HOPE (PHOENIX)

The Good Hope property adjoins the Quartette on the north and consists of four patented claims, one of which is a fraction. Three veins crop out on the Good Hope claim, and the middle vein has been explored by the workings shown in plate 47. The recorded production is very small, but between 1906 and 1911 probably the mine produced some ore that was included with that of the Duplex. It is credited with production in 1913, 1921, and 1933, and lessees were reported to be active in 1934.

The vein has been explored by two shafts, one 210 feet deep and the other about 370 feet deep on the incline, together with 1,350 feet of drifts and 920 feet of crosscuts on four levels, shown in plate 47. Some small stopes have been opened on the 200-foot level, but most of the stoping has been in the vicinity of the air shaft between the 60-foot level and the surface. The strike of the vein is variable but averages N. 70° W. The dip also changes from place to place but is mostly 40°–50° SW. The most conspicuous feature is a nearly vertical fault, which, if interpreted as a normal fault, indicates a drop on the west side of about 100 feet. (See pl. 48.) The country rock is chiefly hornfels, but dikes and masses of andesite porphyry are revealed on most of the levels.

The vein has been explored for a length of 340 feet underground, and pits on the surface indicate a length of 1,000 feet. The vein material near the surface in the vicinity of the air shaft consists of a zone 6 feet wide made up largely of altered rock but contains 26 inches of quartz-cemented breccia on the footwall and 6 inches on the hanging wall. The vein material in the upper workings is stained with copper silicate and iron oxides, but both quartz and copper stain are largely wanting on the 300-foot level. No assays were available.

According to Hewett,[36] the vertical depth to water is 225 feet, and the reported yield 5,000 gallons a day.

The fading out of the vein in the lower workings is not encouraging for prospecting at depth, and further production will probably be limited to small operations near the surface.

The two parallel veins, one on each side of the main vein, have been prospected by shafts, inaccessible in 1931, and open cuts. The dip of the southwesterly vein is 56° SW., and that of the southeasterly vein is 48° SW. Both are narrow quartz veins in hornfels and andesite porphyry.

DUPLEX

The Duplex mine is on the hill south of Searchlight, and the six patented claims and a fraction constituting the property lie largely in the southwest corner of sec. 34, T. 28 S., R. 63 E. It is second only to the Quartette, both in production and in amount of development work. According to an engineer's report, the property was located May 6, 1897, by G. F. Colton as the IXL copper mine, and some ore was shipped to a smelter at Pueblo, Colo. In 1903 it was one of three mines equipped with mills. From 1904 to 1907 it was included with the Good Hope mine as one operation under John Brockman. From 1911 to 1915 it was held by the Homestead Mining & Milling Co. The Duplex Mining Co. operated the property from 1915 to 1918, but subsequent intermittent operations were carried on by lessees until 1930, when the Searchlight Gold Corporation took over the property. This company erected a 100-ton flotation mill, which treated tailings from earlier operations, some ore from the dump, and ore from pillars and small stopes in the mine. Operations by the company ceased in 1931, and the mill was later dismantled. Lessees in 1934 extended level 8 northward to a raise connecting with workings on level 4 from the Searchlight shaft, and the Searchlight shaft was unwatered. No new ore was developed. The mine is credited with production in 1903, from 1906 to 1911, in 1913, and from 1915 to 1934. The total production probably exceeded $650,000 and was derived chiefly from gold, but a considerable yield of lead with silver and copper was obtained, particularly in the later leasing operations.

The underground workings of the mine are accessible from three shafts—the New Years Gift, Fraction, and Searchlight. Most of the later operations were carried on by way of the Fraction shaft. The distribution of the workings that were accessible in 1931 and 1934 is shown in plate 49, and a longitudinal section on the Fraction and New Years Gift veins is shown in figure 24. The Fraction shaft attains a

[36] Hewett, D. F., op. cit. (Water-Supply Paper 578), p. 720.

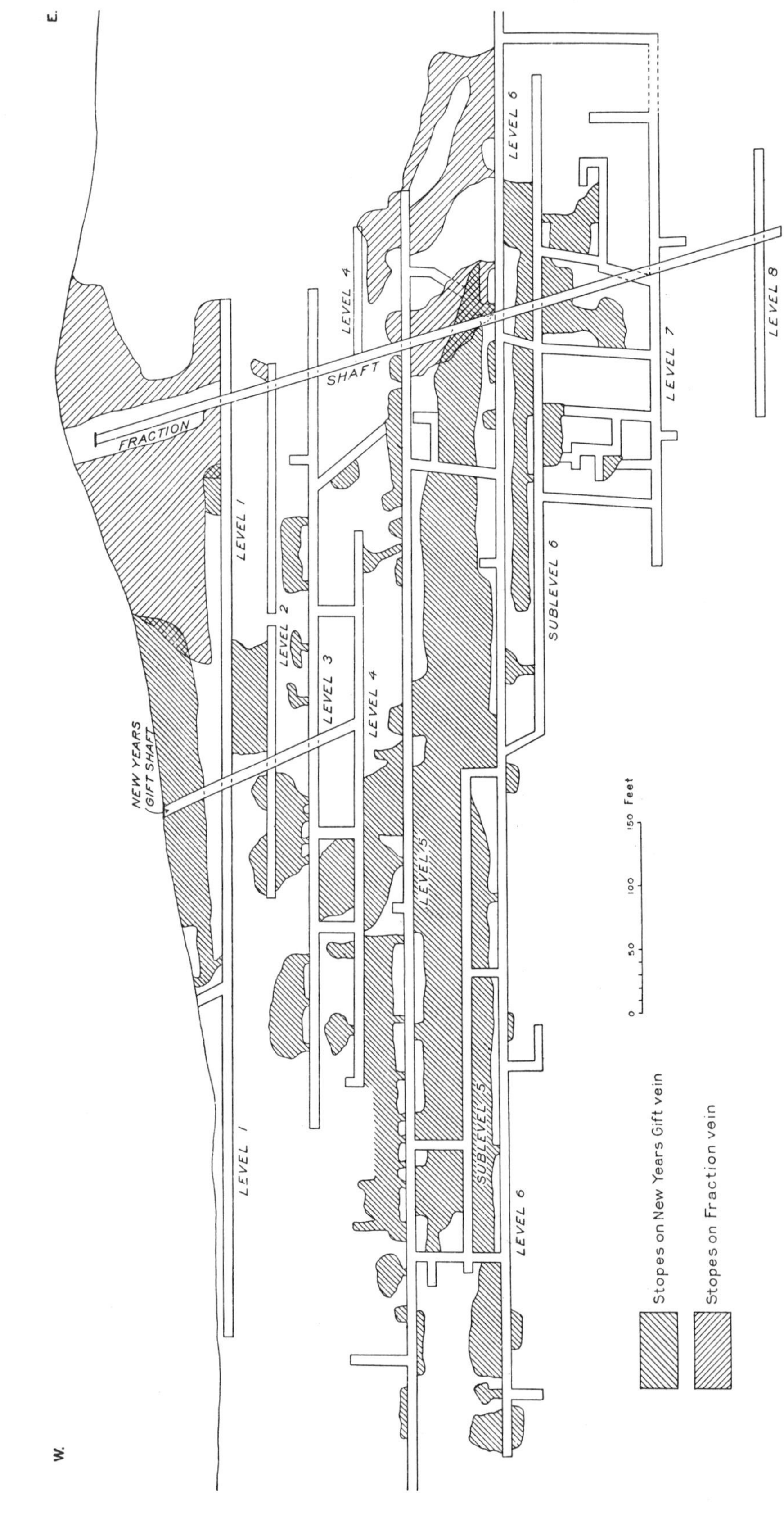

FIGURE 24.—Longitudinal section (vertical projection) of Duplex mine. Furnished by owners. Stoped areas approximate, unsurveyed.

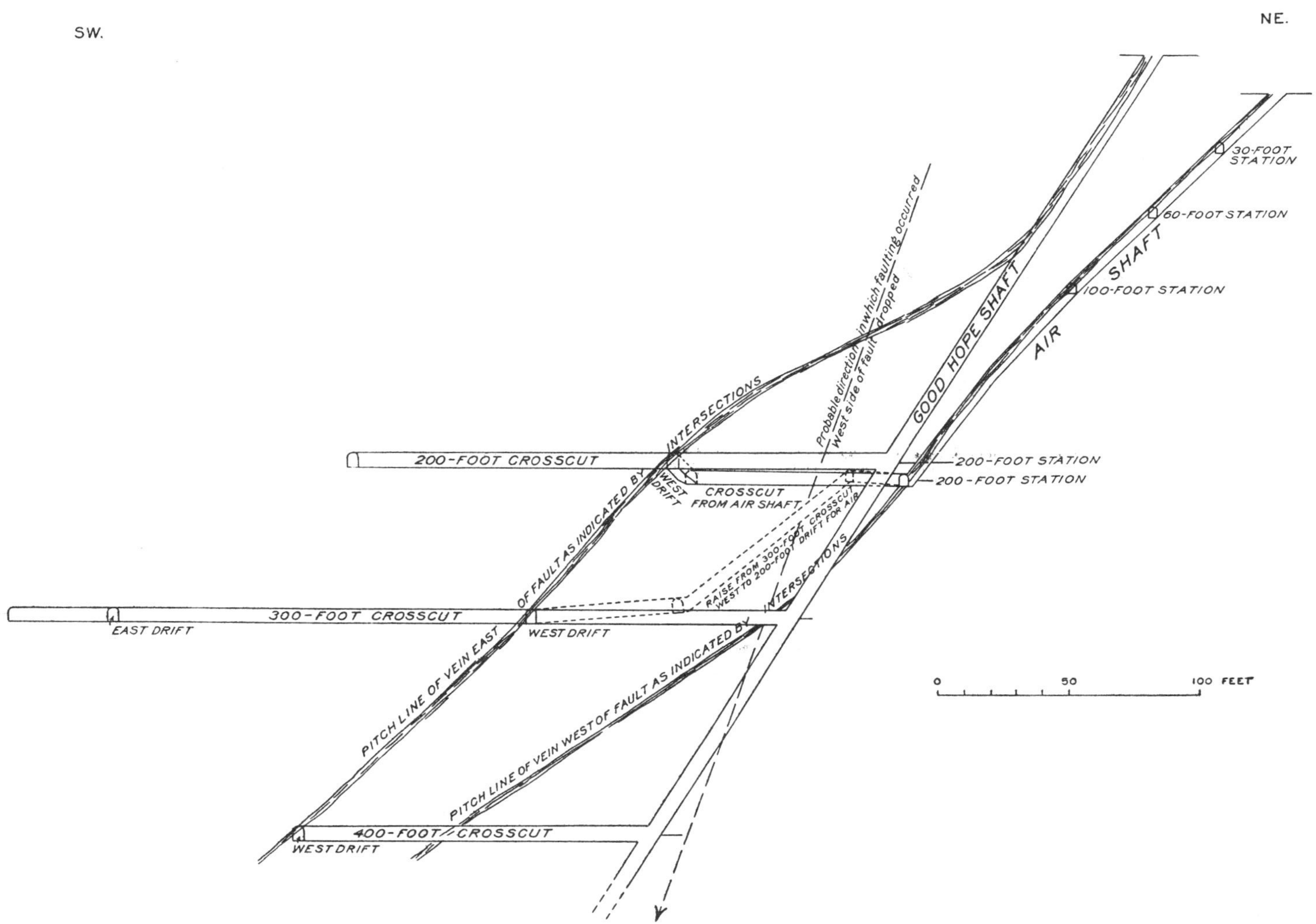

SECTION THROUGH SHAFT OF GOOD HOPE MINE.

Shows relative position of segments of vein on both sides of fault. Furnished by lessee. Segment on north side projected to plane of section.

PLAN OF WORKINGS OF DUPLEX MINE.

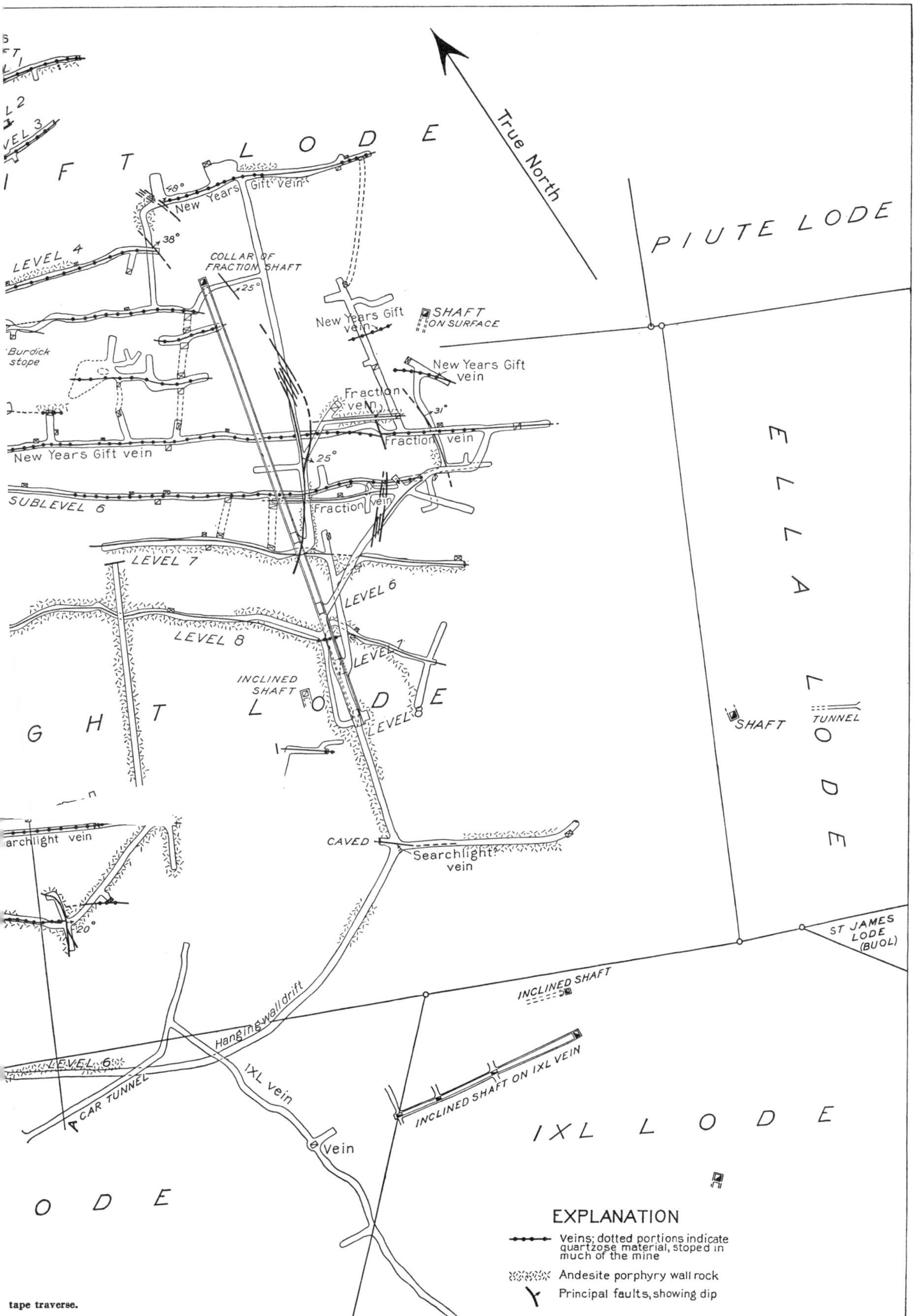

tape traverse.

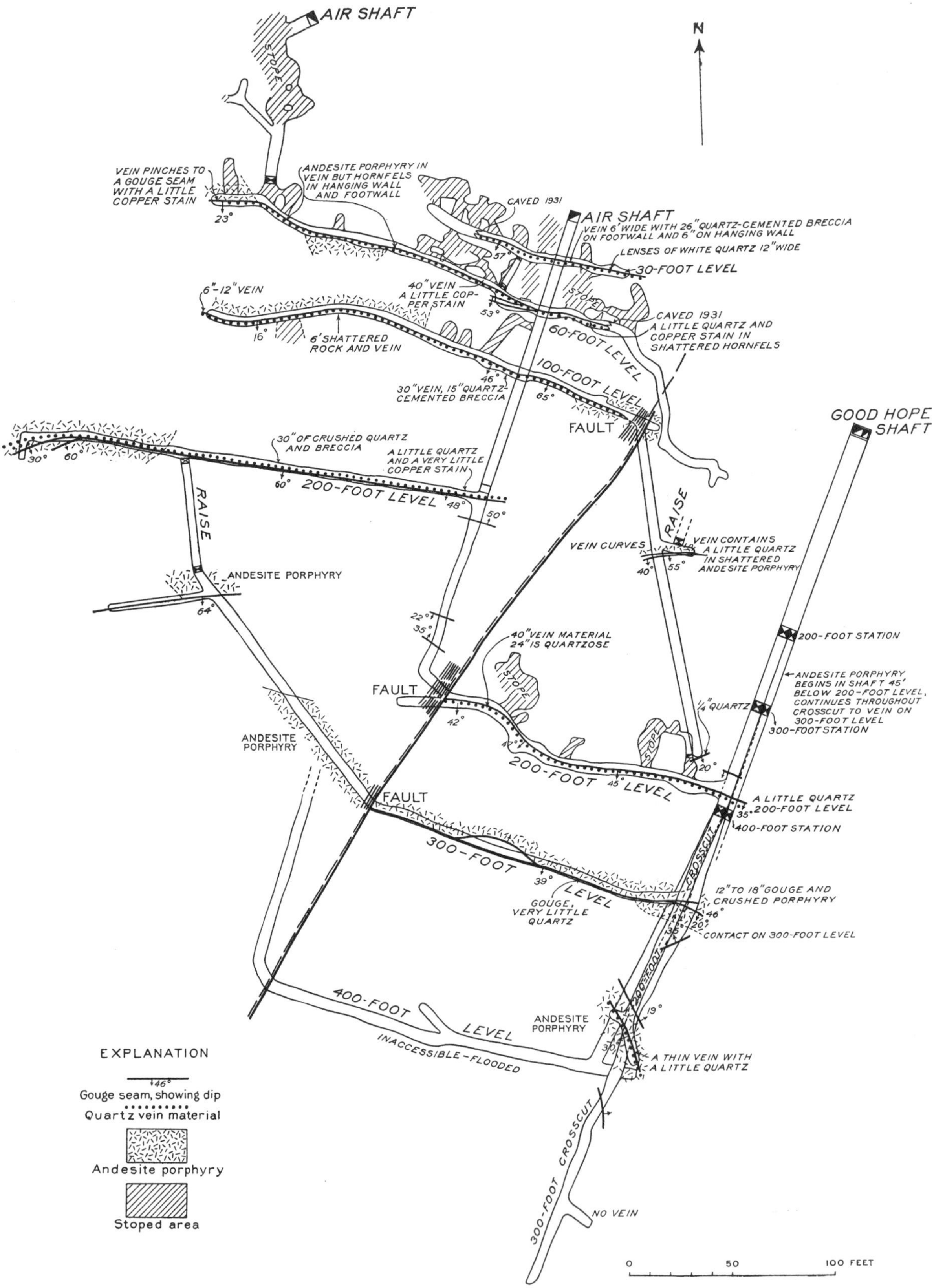

PLAN OF WORKINGS OF GOOD HOPE MINE.

From map made in 1921 and furnished by lessee. Geologic notes added.

depth of 665 feet on the incline or 500 feet vertically below the collar to level 8. The New Years Gift shaft is 265 feet deep on the incline, or 175 feet vertically below the collar, and level 4 from the Searchlight shaft is about 400 feet below the surface.

The country rock is chiefly hornfels, some of which reveals a fragmental structure indicating that it was formerly a tuff or volcanic breccia. Both steeply dipping and gently dipping dikes and irregular masses of andesite porphyry appear in many places in the mine, and their location is indicated on the map (pl. 49). In places lenses of andesite porphyry, apparently fault horses, occur within the vein. The ground has stood up very well in the mine, though levels 7 and 8 from the Fraction shaft and level 4 from the Searchlight shaft are moist or under water, unless kept dry by pumping or bailing. Only occasional bailing has been found necessary to keep the Fraction shaft free of water, but a much larger flow enters the Searchlight shaft.

Four veins, the New Years Gift, Fraction, Searchlight, and IXL, have been explored by the workings shown in plate 49. Of these the New Years Gift has been by far the most productive. Stopes on the Fraction and New Years Gift veins are shown in the longitudinal section (fig. 24). An ore body was removed from the Searchlight vein, but the IXL does not appear to have been productive. The strikes of the veins are variable, but the average for the New Years Gift, as revealed on the longest drifts, is about N. 57° W., that for the Fraction nearly the same, and the strike of the Searchlight vein is about N. 60° W. The IXL vein dips gently, so that its strike is highly variable, but as indicated in the drift in the car tunnel the strike is N. 8° W., and the dip is 10°–25° W. The gentle dip causes the outcrop to curve around the hill as shown by the map (pl. 41). The section (fig. 25) shows the variations in dip, particularly of the New Years Gift vein, which averages about 32° W. through the productive part but steepens to 47° near its junction with the Fraction vein. On level 8 the New Years Gift vein dips 53° W. near the shaft and 70° W. at the north end of the drift. The Fraction vein has an average dip of about 60° W. but is nearly vertical near its junction with the New Years Gift vein. The dip of the Searchlight vein averages from 70°–80° W.

Accessible workings from the Searchlight shaft reveal an ore shoot reaching from level 1 to level 3, which was stoped for a length of about 160 feet. The total vertical range is not known, but the stopes probably extend from the surface to and a short distance below level 3. The vein is not followed by the shaft all the way to level 4 and is probably offset by the fault exposed on that level southeast of the shaft. The stope widths are mostly about 3 feet. Pillars indicate at least 16 inches of quartzose vein material in the ore shoot, but on level 4 the vein consists of only narrow quartz stringers. Nodules of galena were

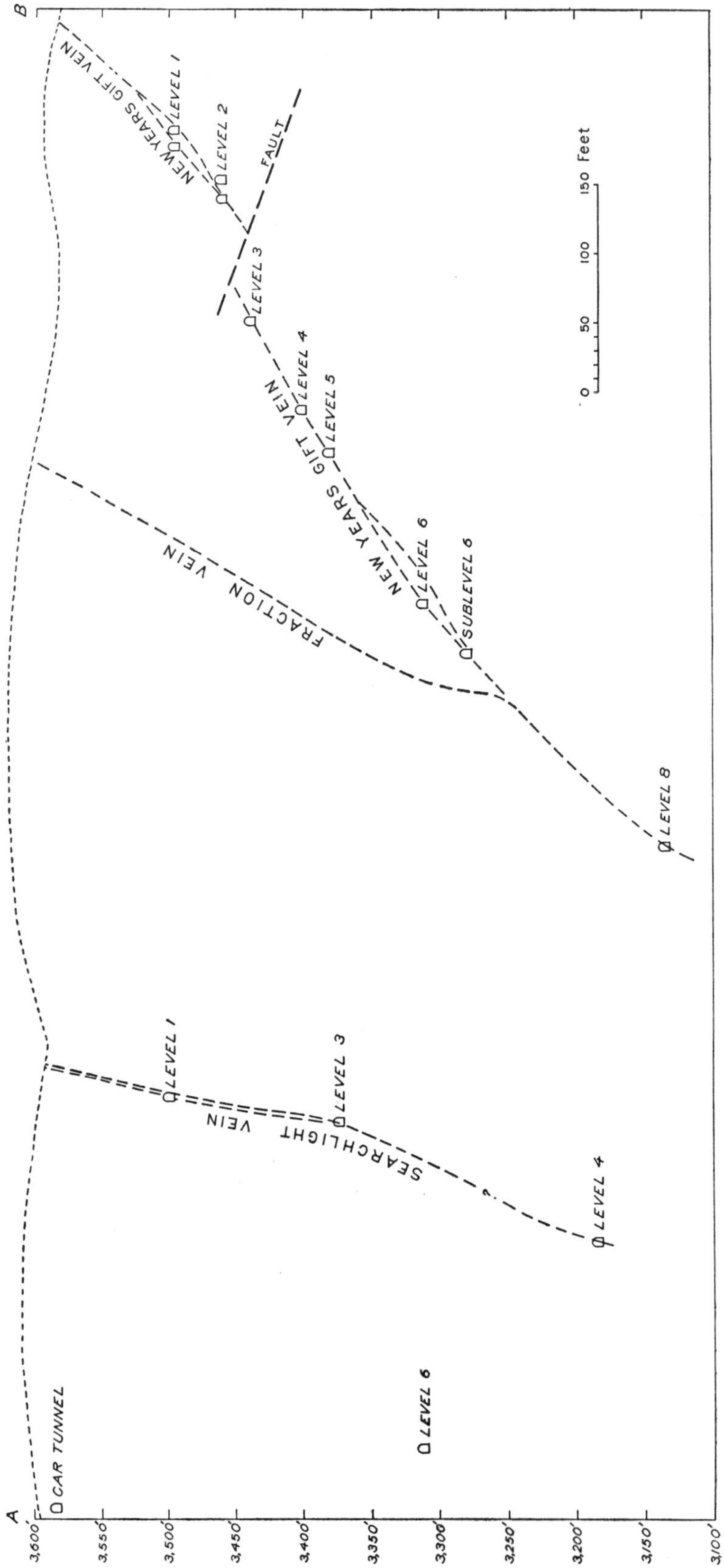

FIGURE 25.—Section through Duplex mine showing relations of veins. Assumed datum.

obtained from a small ore shoot within the footwall of the vein on level 1.

The inclined shaft on the IXL vein reveals a little quartz-cemented breccia and copper stain, but no ore is known to have been mined from this vein.

The New Years Gift vein was productive from a point a short distance above its junction with the Fraction vein to the surface, but most of the stoping appears to have been done on a long, narrow ore body lying mostly between levels 5 and 6. The longitudinal section (fig. 24) indicates a pitch of about 10° E. for this ore body or group of ore bodies. The structure of the vein is complex, and workings reveal intersecting and splitting gouge seams that enclose lenses of country rock and vein matter as much as 15 or 20 feet thick. The ore appears to have occurred in lenses separated by gouge seams or by masses of rock or barren vein material. Some of the stopes are 10 feet or more wide, though the usual width is 4 or 5 feet. The vein material consists of a breccia of country rock (pl. 44) cemented with quartz, which is generally stained and partly replaced by copper minerals, chiefly chrysocolla, and iron oxides. Lumps of cerusite and some large bodies of cerusite occurred in this material. A few lumps of coarse galena are preserved in the ore. All the lead carbonate was valuable ore, but by no means all the siliceous vein matter was of value. The operators in 1931 stated that good ore, other than lead carbonate ore, could be distinguished by assay only. A thin section of greenish and brownish siliceous ore from a pillar at the Burdick stope, on level 5 revealed numerous minute flakes of gold in quartz or in small areas of limonite.

Weathering of vein material has been complete throughout the mine, except for chunks of galena, one of which was found within a few feet of the surface, in the New Years Gift shaft. A specimen of galena was found by the assayer at the Duplex mill to assay in gold and silver $16.94 a ton (at the old price), and the lead carbonate coating assayed $21.49 a ton. In another specimen the galena assayed $1.03 to the ton, whereas the lead carbonate assayed $69.04 to the ton. Zinc minerals other than films of mottramite were not observed in ore specimens from the Duplex mine, but the zinc reported in smelter returns was probably present in the form of the inconspicuous silicate, hemimorphite (calamine).

A table of smelter returns from photostats in an engineer's report is given below. The ore on which the returns are given was mostly handworked or "chlorided" by lessees. This table does not show the total production of the mine, as considerable ore was milled locally. The figures indicate that shipments containing over 20 percent of lead not only contained more gold than the average of all the shipments, but also a higher proportion of silver to gold and a much lower proportion of copper.

Smelter returns on shipments of crude ore from Duplex mine

Date	Weight (pounds)	Gold (ounces per ton)	Silver (ounces per ton)	Lead (percent)	Copper (percent)	Zinc (percent)	Sulphur (percent)	Iron (percent)	Insoluble matter	Gross value per ton
1907	58,783	2.31	11.8	27.2	2.0					
1907	32,769	1.75	11.7	21.6	2.4					
1907	14,897	2.4	6.4	29.7	.3					
July 6, 1915	2,532	.07	4.31	5.0	7.4	1.5	0.6	5.5	69.8	$26.76
Do	15,380	1.49	22.1	35.9	4.3	1.2	2.8	4.6	38.2	77.29
Do	57,666	.54	4.8	7.5	3.82	1.2	.6	3.9	72.2	27.73
Aug. 23, 1915	4,398	2.87	30.7	39.4	3.47	.4	3.4	3.78	32.3	98.72
Do	71,782	.9	4.9	8.6	2.33	1.3	.4	4.8	69.5	29.11
Sept. 8, 1915	51,600	.98	6.9	9.6	3.43	.5	.5	4.4	67.8	34.61
July 22, 1916		1.97	14.43	24.4	4.0					77.26
Do		1.31	5.49	5.6	4.1					42.71
Aug. 29, 1916	3,960	1.32	6.51	5.0	3.7					43.77
May 19, 1917	18,390	1.64	4.3	7.4						41.32
Do	45,658	1.94	5.9	5.4	3.3					54.17
June 7, 1917	15,739	1.9	5.34	5.6	2.8					51.85
Do	2,670	3.07	13.33	26.8						100.80
June —, 1917	54,356	1.48	2.56	6.6						36.58
June 29, 1917	69,956	1.74	2.81		2.1					38.88
July 18, 1917	67,866	.965	3.66	9.8	.7			4.1	72.5	32.69
Aug. 12, 1917	70,430	.685	2.86							15.14
Do	34,559	1.04	2.36	1.2						32.26
Sept. 6, 1917	34,304	2.45	3.41		2.1					52.58
Do	45,011	.93	4.62	11.8	.6					40.97
Oct. 20, 1917	65,498	1.02	2.98		2.1					24.72
Nov. 23, 1917	10,980	2.46	8.6	6.7	5.77	1.8	.4	4.6	67.9	76.24
Do	12,280	.21	4.7	8.0	2.23		.4	5.0	69.4	19.86
Feb. 6, 1918	38,338	1.94	2.8	8.0	1.15	2.4	.8	3.4	70.4	47.27
Do	36,292	2.81	4.1	3.5	2.57		.6	5.2	76.9	62.55
Do	5,540	1.705	15.9	58.9	.3		.2	3.5	12.4	101.14
Mar. 9, 1918	89,332	3.22	5.55		2.98			5.5	75.4	73.96
May 25, 1918	9,334	4.45	3.70	9.8	1.65	1.8	.2	5.0	67.2	100.15
Do	51,486	4.08	3.20		2.56			5.5	76.5	87.07
July 17, 1918	37,946	1.20	4.5	2.0	2.86			6.4	77.1	35.27
Oct. 30, 1918	3,952	4.19	4.08	7.0						89.22
Do	57,953	2.005	6.1	13.4	2.4					61.11
Nov. 6, 1918	70,219	5.96	7.07	9.2	4.4					140.67
Dec. 23, 1918	69,539	3.38	4.8	7.4	2.9					76.89
Feb. 4, 1919	35,068	2.02	7.3	14.8	2.7					57.38
Feb. 10, 1920	62,952	4.33	6.0	7.2	2.8		.4	5.9	71.6	102.32
Mar. 15, 1920	59,270	4.68	5.4	7.1	3.35		.8	6.4	70.0	109.56
May 11, 1920	60,720	6.0	6.1	9.5	3.0		.6	6.3	69.9	137.30
June 2, 1920	65,668	5.375	6.4	8.5	3.45		.8	6.1	69.0	123.89
June 18, 1920	68,754	6.19	7.3	6.3	2.7		.6	5.6	71.7	136.62
July 12, 1920	75,836	9.3	10.0	13.2	4.4		.6	5.4	63.9	209.14
Aug. 7, 1920	60,302	6.18	6.8	7.8	3.8		.8	5.4	69.5	140.10
Aug. 13, 1920	62,594	6.49	8.8	9.3	4.3		1.0	5.6	67.6	150.93
Aug. 26, 1920	68,754	6.17	7.3	8.3	2.7		.6	5.6	71.7	137.95
Aug. 27, 1920	67,760	5.74	9.3	10.4	4.37		.6	5.9	58.34	139.07
Sept. 15, 1920	62,300	5.68	8.5	7.6	5.5		.8	5.4	63.5	136.30
Sept. 26, 1920	59,450	4.32	5.6	7.2	3.4		.6	6.4	70.8	101.41
Oct. 2, 1920	66,936	7.23	11.5	6.9	6.4		1.0	5.1	65.8	169.12
Nov. 6, 1920	64,576	3.35	8.0	3.8	5.0		1.0	5.0	72.2	79.41
Nov. 18, 1920	64,024	3.11	4.4	2.2	3.75		.8	4.1	76.6	67.39
Nov. 27, 1920	62,210	4.41	6.9	5.2	4.7		1.0	5.1	39.2	98.70
Dec. 3, 1920	64,576	3.35	8.0	3.8	5.0	1.5	1.0	5.0	72.2	80.69
Dec. 9, 1920	63,440	3.26	5.2	4.3	3.9		1.0	5.0	73.6	71.67
Jan. 5, 1921	59,616	6.22	7.4	4.2	5.95	1.0	.08	3.5	68.0	132.37
Do	67,382	7.58	11.9	7.9	5.5	1.8	1.0	5.5	64.8	164.14
Jan. 17, 1921	68,220	6.25	7.8	4.2	4.45		1.0	3.5	74.6	131.64
Jan. 19, 1921	60,298	4.12	6.4	7.3	2.15		1.4	2.8	70.8	89.39
Jan. 28, 1921	76,968	3.56	5.6	4.5	2.4		1.4	2.5	69.8	76.76
Feb. 1, 1921	61,690	3.13	4.1	6.2	2.4		.6	5.5	72.8	67.94
Feb. 9, 1921	60,784	3.28	5.9	2.35				3.9	70.2	72.34
Feb. 10, 1921	87,404	4.14	7.4	6.5	2.6		1.2	3.5	70.8	90.70
Mar. 16, 1921	89,984	2.96	8.0	5.0	2.9		1.0	3.4	74.8	68.42
Apr. 7, 1921	74,726	5.08	5.7	7.8	2.0		1.0	5.1	73.2	169.00
Do	77,722	3.43	6.3	4.0	3.3		.8	4.8	74.0	76.14
Apr. 28, 1921	65,606	6.5	9.4	8.2	3.0		1.0	4.5	70.8	138.90
June 4, 1921	68,488	3.22	7.8	7.6	3.84	1.05	.4	4.0	71.5	76.38
June 17, 1921	60,836	7.155	7.8	4.8	2.4	.62	1.2	3.8	76.8	147.22
Aug. 13, 1921	75,724	4.77	5.7	3.5	2.3	.55	.7	3.7	78.7	99.83
Sept. 28, 1921	74,438	4.655	4.48	4.4	1.79	.7	.3	3.7	79.0	98.22
Oct. 1, 1921	67,724	4.49	6.01	4.65	3.88	1.45	.7	2.6	75.2	99.08
Nov. 1, 1921	75,076	4.205	7.03	7.55	4.23	.65	.6	4.6	71.1	97.09
Nov. 17, 1921	64,870	5.625	7.6	11.9	2.69	1.65	.55	4.3	66.6	125.57
Dec. 28, 1921	69,618	3.39	4.35	4.65	3.4			4.6	76.4	76.82
Feb. 3, 1922	79,618	2.96	6.25	8.9	3.47	.8	.4	4.5	71.5	70.71
Apr. 6, 1922	74,656	3.25	6.05	7.55	3.97		.4	5.0	71.6	77.22
July 5, 1922	64,044	2.69	7.03	7.4	3.96	.8	.4	5.8	70.0	67.62
Do	22,210	3.755	8.9	14.7	3.89	.6	.8	4.6	62.6	96.76
July 19, 1922	60,348	6.17	4.85	6.0	2.79			4.4	74.8	132.04
Sept. 27, 1922	81,234	3.145	7.05	6.95	3.03	.4	.3	4.7	71.6	76.43
Oct. 26, 1922	72,897	1.82	2.9	1.7	1.7	.4	.4	4.8	78.9	38.42

Smelter returns on shipments of crude ore from Duplex mine—Continued

Date	Weight (pounds)	Gold (ounces per ton)	Silver (ounces per ton)	Lead (percent)	Copper (percent)	Zinc (percent)	Sulphur (percent)	Iron (percent)	Insoluble matter	Gross value per ton
Nov. 7, 1922	74,942	2.22	5.6	7.05	3.2	1.6	0.4	4.75	71.5	$57.08
Nov. 16, 1922	74,822	1.765	3.45	2.75	1.93	--------	.4	4.4	79.7	38.22
Jan. 16, 1923	81,568	3.055	6.68	8.15	3.08	.9	.3	4.3	71.5	77.60
Apr. 13, 1923	76,596	2.04	4.63	6.9	3.48	2.0	.3	4.25	70.5	56.71
June 5, 1923	42,612	3.04	5.39	8.3	3.43	--------	--------	4.75	70.7	77.18
June 8, 1923	27,352	2.375	3.75	5.25	3.19	--------	--------	4.4	72.3	57.94
July 30, 1923	50,398	2.69	7.4	5.07	5.12	--------	--------	5.7	70.4	66.56
Nov. 21, 1923	53,742	2.3	8.15	9.0	4.84	--------	--------	6.05	65.75	69.87
Do	65,614	4.015	9.75	10.55	4.36	--------	.8	3.75	68.5	96.33
Do	4,632	3.26	24.25	50.0	1.85	--------	--------	7.05	20.2	122.80
Apr. 24, 1924	67,664	2.33	2.55	2.4	3.2	.4	.2	5.0	75.2	48.60
June 11, 1924	55,468	2.69	8.0	8.4	6.15	.3	.3	3.25	67.1	70.69
July 2, 1924	65,156	2.51	5.85	4.9	3.11	.7	--------	5.2	70.25	59.50
July 11, 1924	23,148	3.925	22.45	37.8	1.6	3.4	--------	9.05	25.3	128.15
Sept. 30, 1924	79,870	4.065	5.95	11.05	2.01	--------	--------	3.35	68.5	97.20
Oct. 29, 1924	29,120	2.625	2.4	4.7	1.54	.8	.4	3.3	79.0	57.39
Do	8,612	2.3	16.0	40.15	.85	2.2	1.0	5.75	30.7	101.11
Jan. 6, 1925	33,728	4.085	4.93	11.5	1.62	--------	--------	2.7	72.5	99.81
Do	10,330	4.46	14.5	51.65	.57	--------	--------	4.6	18.7	168.72
Do	8,692	5.58	3.9	10.5	.43	--------	--------	3.5	73.2	125.83
Feb. 10, 1925	77,194	6.51	5.4	10.9	1.59	4.0	.2	3.3	71.2	147.57
Apr. 6, 1925	5,052	3.69	5.9	22.9	.45	1.2	.4	5.85	55.05	105.44
Do	68,860	4.51	4.0	5.9	1.99	--------	.6	2.8	76.97	99.54
May 22, 1925	19,002	2.07	8.15	30.15	.6	.25	1.2	.62	42.65	77.50
Do	63,082	6.55	5.7	11.5	3.64	2.7	2.3	3.7	64.0	149.68
June 10, 1925	64,566	5.81	5.85	8.85	3.53	4.4	.8	4.2	64.75	131.37
July 21, 1925	9,496	2.0375	6.3	23.35	.58	2.3	.7	5.3	51.10	69.44
Do	84,504	6.32	5.45	8.35	3.26	2.8	.8	4.3	67.0	139.62
Aug. 18, 1925	12,812	2.57	6.25	30.6	.69	2.1	.8	5.0	42.95	90.54
Do	78,272	7.17	9.8	15.15	3.82	2.35	.85	3.75	59.65	169.55
Sept. 15, 1925	11,402	1.83	3.45	15.65	.58	1.45	.6	5.05	62.15	58.92
Do	73,266	7.735	7.05	11.2	4.37	2.95	.4	4.15	62.8	176.89
Oct 23, 1925	5,262	1.955	3.7	16.95	.5	1.45	.2	3.8	61.4	63.24
Do	58,268	5.82	4.2	6.05	3.8	1.9	.3	3.8	70.15	129.14
Nov. 24, 1925	10,230	8.56	6.0	16.55	.54	1.45	.3	5.6	57.50	194.73
Do	83,198	4.155	3.38	5.2	2.33	1.45	.25	3.9	73.7	92.79
Jan. 26, 1926	62,964	8.205	8.48	14.8	2.5	3.6	1.2	4.0	60.65	187.81
Do	9,006	11.03	10.75	35.35	.75	3.1	1.4	3.9	36.40	270.02
Feb. 19, 1926	54,720	4.93	4.95	12.0	2.05	2.95	--------	3.85	62.8	117.02
Mar. 18, 1926	69,878	6.965	6.25	15.5	1.91	2.4	.2	3.7	62.3	161.30
Apr. 13, 1926	9,570	10.35	9.5	45.55	.71	--------	.6	3.15	27.45	261.04
Do	89,140	6.79	6.25	9.65	2.53	4.85	.15	4.8	61.9	149.78
May 13, 1926	77,890	5.895	4.05	10.3	2.42	3.0	.5	4.45	65.75	130.82
June 18, 1926	11,366	9.26	9.13	33.65	.93	1.5	.8	4.5	39.9	222.32
Do	58,252	5.4025	5.95	8.7	2.44	1.2	.6	4.45	69.6	120.40
Aug. 5, 1926	8,518	9.575	7.25	31.2	1.06	.3	.7	3.9	42.0	229.28
Aug. 4, 1926	15,483	4.12	3.35	8.95	1.82	.7	.6	3.95	72.4	94.39

Average metal content per shipment

Gold	ounces per ton	3.97
Silver	do	6.92
Lead	percent	11.77
Copper	do	2.90
Zinc	do	1.63

Average metal content per shipment of lots containing over 20 percent lead

Gold	ounces per ton	4.45
Silver	do	13.91
Lead	percent	36.34
Copper	do	1.42
Zinc	do	1.63

The possibilities of finding ore shoots west of the fault in the Duplex mine appear to have been exhausted by the exploratory work already done. It does not seem that exploration east of the fault has been complete, especially as the normal fault movement should have depressed the productive part of the vein below its position in the workings to the west. Any further production of the mine will probably be limited to this area.

CYRUS NOBLE

The Cyrus Noble mine and mill are about 1,500 feet west of the Duplex mine, and the claims are in the southwestern part of sec. 34, T. 28 S., R. 63 E. According to Ransome,[37] the mill at the Cyrus Noble had recently been erected at the time of his visit in 1906, and

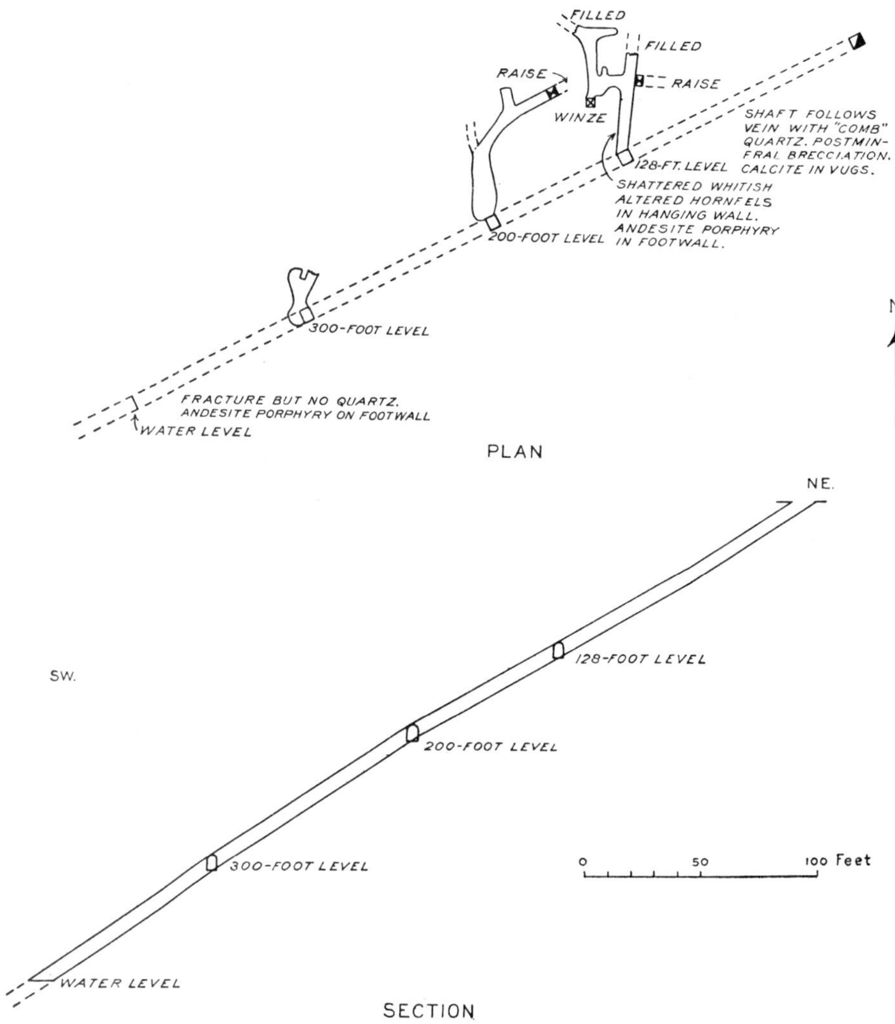

FIGURE 26.—Plan and section along shaft, Cyrus Noble mine. From compass and tape traverse.

preparations were under way for hoisting ore. The mine is credited with a small production in 1904–8, 1914, 1917–23, and 1928. The mill has been operated from time to time as a custom mill, and ore from other mines was probably included in production figures with that of the Cyrus Noble. According to Mineral Resources of the United States for 1908, the mine had 1,500 feet of drifts on various levels. The map (fig. 26) shows that only a small part of these workings were accessible in 1931.

[37] Ransome, F. L., op. cit. 75.

The shaft extends 385 feet to water level and an unknown distance below. The average angle of slope is 30° S. 65° W., and the vertical depth to water is about 195 feet. As shown by figure 26, three levels extend northward from the shaft. The shaft and the drifts reveal a gently dipping vein with comb quartz and brecciated country rock, chiefly hornfels, though andesite porphyry appears in several places in the footwall. The vein curves sharply, but the average strike appears to be about due north. The shaft follows the vein fracture,

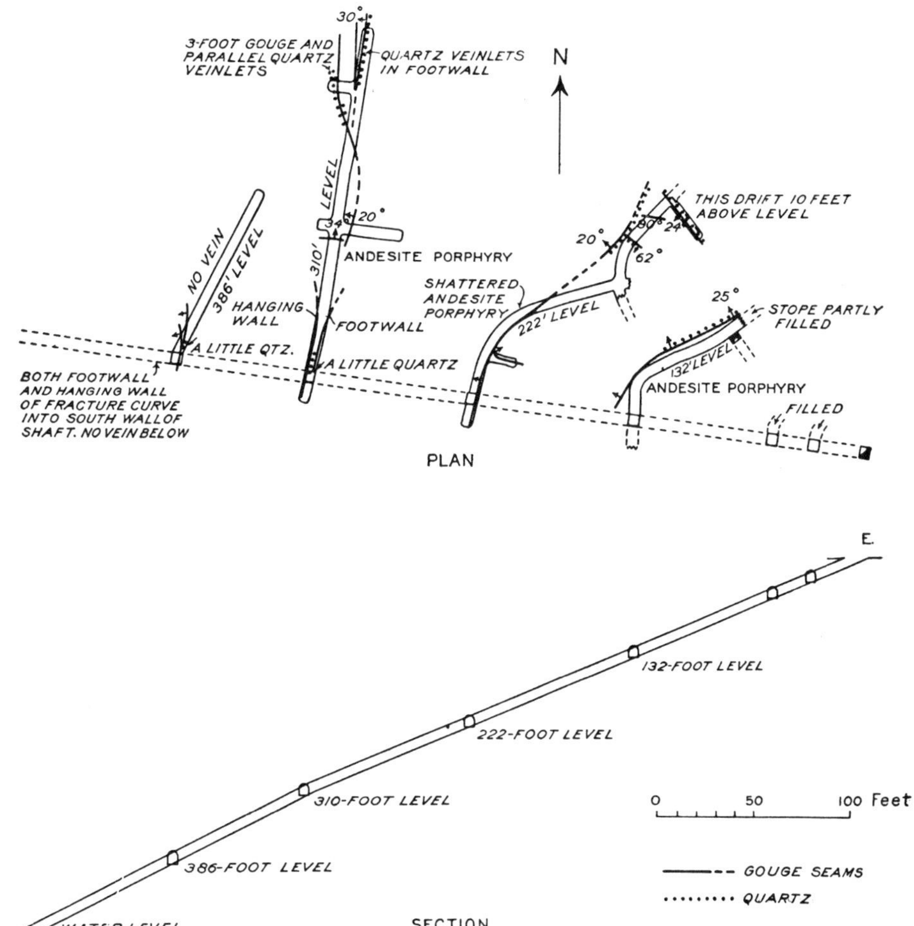

FIGURE 27.—Plan and section through shaft, Searchlight Parallel mine (Elvira ? shaft). From compass and tape traverse.

but no quartz was seen near the water level. The mine is reported to have had an unusually high proportion of silver for the district, and the records show about 1 ounce of gold to 9 ounces of silver in the first years of production.

SEARCHLIGHT PARALLEL

The Searchlight Parallel property is on the west side of the district in the northwestern part of sec. 34, T. 28 S., R. 63 E. The workings shown in figure 27 are about 900 feet north of the Nipton road. There is no surface establishment. According to Mineral Resources of the United States the mine was productive in 1905, 1908, 1911, 1913, 1915, and 1917–18, but the total output has been small. By 1907 the shaft was 492 feet deep on the incline, and 1,900 feet of drifts were reported. In 1908 ore was sent to the Cyrus Noble mill.

The shaft has an average slope of 23° N. 82° W. and follows the vein to the 386-foot level. The water level in 1931 was about 460 feet down the shaft, or about 180 feet below the surface. As surface flood waters have access to the shaft, the water level may fluctuate over a wide range. In 1931 the workings on four levels were accessible to the extent shown in figure 27. On account of the low dip of 20° to 25°, curvature of the vein causes a wide variation in strike. Probably the average strike is about north, as indicated on the 310-foot level, though the upper workings indicate a strike of about N. 60° E.

The vein consists of quartz and shattered altered country rock that is chiefly andesite porphyry. Postmineral brecciation is evident, and in most places the vein material consists of quartz veinlets and gouge seams. A maximum of 1 foot of quartz was seen 55 feet from the surface associated with 2 feet of partly silicified altered country rock. Such stopes as are visible are small and very irregular in shape. Weathering of vein material has been complete throughout the workings.

SANTA FE

The Santa Fe shaft has been important as a source of water. It was being utilized in 1931 by the Duplex mine, and, according to Hewett,[38] it was the source of water for Searchlight in 1922. The shaft is near the highway north of Searchlight, near the north line of sec. 34, T. 28 S., R. 63 E. The patented claims constituting the property extend northward in sec. 27. Probably most of the work was done prior to 1906. According to Mineral Resources of the United States for 1908, the shaft was 245 feet deep, and there were 2,000 feet of drifts and laterals. There is no record of production.

The shaft is inclined 47° W. and extends 198 feet to water level (145 feet vertically). The main level shown in figure 28 is 158 feet from the surface. It consists of a drift 210 feet long on a vein striking N. 10° E. on the average and dipping 45°–55° W. and a long crosscut to the east that reveals several fractures, two of which contain a little quartz. Quartz veinlets with gouge seams are exposed for about 100 feet along the drift. No assays of this material were available. The country rock throughout this level is fine-grained andesite hornfels.

SOUTHERN NEVADA (SPOKANE)

The Spokane shaft of the Southern Nevada mine is near the highway about 3,500 feet north of the Santa Fe. Most of the original

[38] Hewett, D. F., op. cit. (Water-Supply Paper 578), p. 720.

property is in the northeastern part of sec. 27, T. 28 S., R. 63 E. The underground workings were not examined in 1931. According to Ransome [39] water was reached in the shaft 200 feet below the surface.

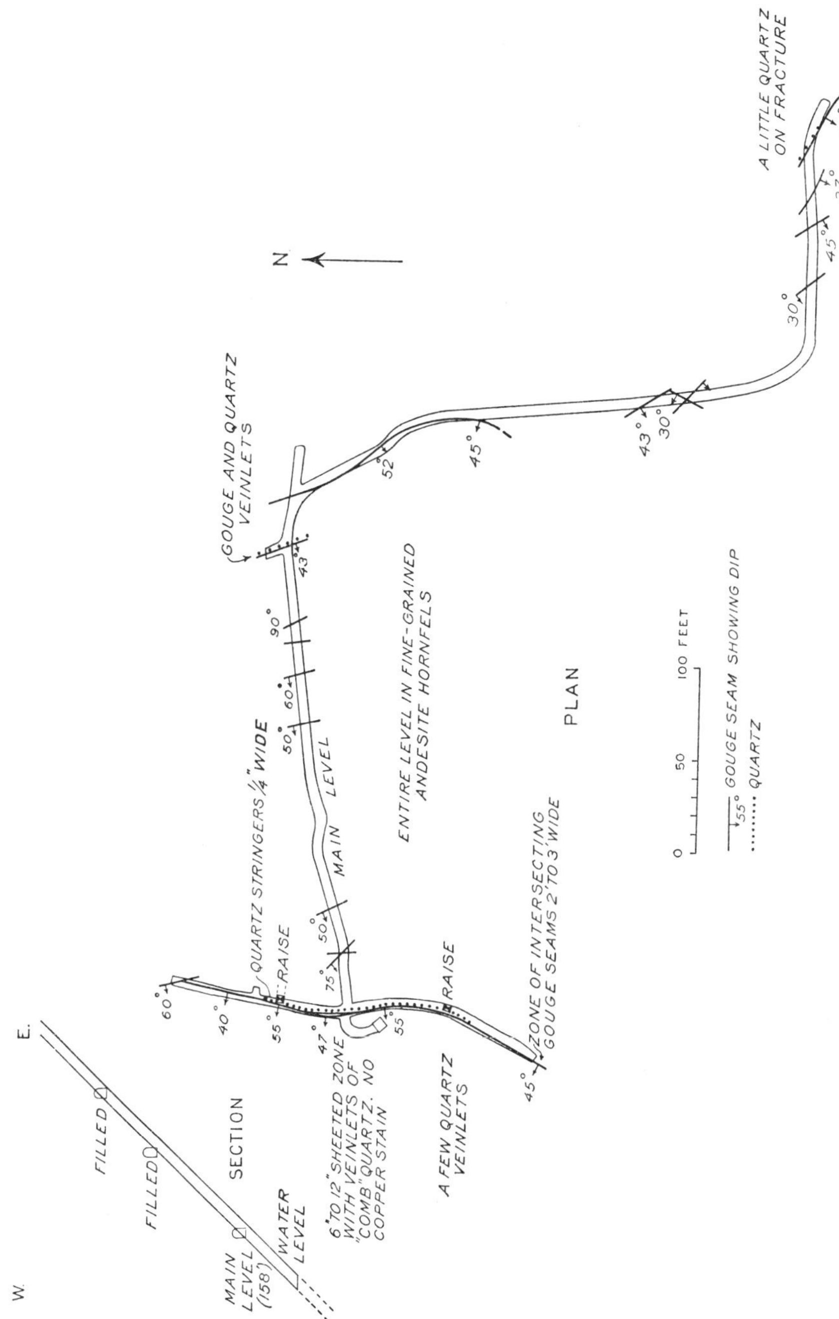

FIGURE 28.—Plan of main level and section along shaft, Santa Fe mine. From compass and tape traverse.

Though extended but 4 feet farther, the shaft supplied ample water for the 10-stamp mill. Some good ore was said to have been found, but development ceased after the discovery of the Blossom deposit, 4,000 feet to the north. The Blossom ore was milled here.

[39] Ransome, F. L., op. cit., pp. 72, 75.

A vein exposed on the surface is explored by three shafts or deep pits, two of which are very close to the road south of the site of the mill. The vein strikes N. 70°–80° W. and dips 80° S. Vein material on the dump consists of a breccia of country rock cemented with comb quartz and containing coarse calcite in vugs and a little copper stain. Some lamellar calcite and quartz also occur on the dump. North of the mill two shafts inclined to the south were inaccessible in 1931.

SEARCHLIGHT M. & M.

The Searchlight M. & M. or Peerless mine is in the northern part of the district, west of the highway and slightly more than 1½ miles north of Searchlight. The property lies partly in sec. 27 and partly in sec. 22, T. 28 S., R. 63 E. The mine has a continuous record of production from 1905 to 1923, and produced in 1926, 1933, and 1934. The total production has not been large, however. There is an extensive surface establishment, comprising headframe and hoisthouse, mill, shops, and numerous houses. The operators in 1934 were exploring above water level and milling some low-grade material. The property was visited only briefly in 1934. The map (pl. 50) is taken from an old print furnished by Mr. A. S. Gaines and was not checked. Most of the workings shown were inaccessible, particularly the extensive 300-foot level, which is flooded. According to the operators in 1934, the distance down the shaft to the 250-foot level is 283 feet. Hewett [40] states that the mine is reported to yield 350,000 gallons of water a day.

As indicated by plate 50, the strike of the vein averages about N. 75° W. The dip is about 70° S. The shaft is steeper and is in the footwall of the vein on the lower levels. On the 250-foot level east of the shaft the country rock is andesite porphyry on both walls. The vein material, consisting of quartz and silicified country rock, is 6 feet wide in one place, and quartz veinlets extend 15 feet into the hanging wall. This material was reported to be of very low grade. In places casts of lamellar calcite are preserved in the quartz. Narrow stopes above this level appear to be extensive. All the vein material that was seen is weathered.

BLOSSOM

The Blossom mine is unique in that the major part of its production came from a nearly flat or saucer-shaped ore body that cropped out on all sides of a low hill. The vein in the deeper workings was also nearly flat in places and highly variable in strike. The mine is east of the highway in the northern part of the district between the Searchlight M. & M. and Pompeii mines. The patented claims are in the southeast-central part of sec. 23, T. 28 S., R. 63 E. Production

[40] Hewett, D. F., op. cit. (Water-Supply Paper 578), p. 720.

is recorded for the years 1903–8, 1913–18, 1924, and 1934. The property was held by the Southern Nevada Mining & Milling Co. and was prospected and developed shortly after the completion of the Spokane shaft. The ore was milled at the Southern Nevada mill. The bulk of the production, about $325,000, was obtained prior to Ransome's visit in 1906.[41] At that time the surface ore body had been worked out, and explorations were being carried on to find additional ore. The two deep shafts had been sunk and levels run. In 1934 a group of lessees was active, and a body of ore was being taken out at the west end of level 2.

The possibility was entertained that such a flat vein might be part of the sole of a thrust fault, so the map (pl. 51) showing the areal geology was made, as well as a map (pl. 52) showing the principal accessible underground workings. Stopes extend under most of the area outlined as the surface ore body, and the hanging wall is supported by pillars. Workings on three levels shown in plate 52 are accessible from the main shaft at the southeast side of the surface ore body, and workings on two levels are accessible from the eastern or "fault" shaft. One of the levels from the "fault" shaft is shown in figure 19, but Ransome [42] states that drifts were run on the 130-, 230-, and 330-foot levels, and the inclined shaft was 430 feet long. The upper part of this shaft is inclined 24° E., but the lower part is inclined 43°. Level 1 is about 75 feet vertically below the collar of the main shaft, level 2 is about 95 feet below the collar, and level 3 is about 137 feet below the collar. The level shown on the "fault" shaft is at practically the same altitude as level 2.

As shown in plate 51, the Blossom mine is at the contact of a sill-like protuberance of the quartz monzonite body. The invaded rocks include both andesite hornfels and the andesite porphyry that is intrusive into the hornfels. Some of the andesite hornfels is porphyritic and so closely resembles finely porphyritic facies of the andesite porphyry that they are not readily distinguished in the field. Furthermore, some of the quartz monzonite at the contact very closely resembles the andesite porphyry. Of primary interest is the fact that the contacts are not appreciably offset by the footwall and breccia zone of the surface ore body. This indicates that conditions on a curving, nearly horizontal surface favored extensive brecciation with very little relative movement. The surface body appears to curve gently downward toward the conspicuous gouge seam or fault followed by the "fault" shaft but may not have actually reached the fault. The vein as exposed on level 1 of the main shaft curves toward the northeast and may in its northeastward extension merge with the breccia of the surface ore body. West of the main shaft the surface ore body has a rather distinct footwall which dips

[41] Ransome, F. L., op. cit., p. 72.

[42] Ransome, F. L., op. cit., p. 73.

18° N., but it fades out on the north side. The relation of the surface ore body to the vein below west of the main shaft is shown by the section in figure 29. The west shaft is vertical for 28 feet and inclined 41°–45° S. on the vein exposed in the lower workings. It is 17 feet from the bottom of the surface ore body to the lower vein, which curves gently under it. The large gouge seam explored by the "fault" shaft and the levels driven from it has an average strike of about N. 13° E. and dips 36°–43° E. Quartz was seen on this fault in only one place, and it appears to be largely if not wholly a postmineral fault. It may have offset the mineralized breccia of the surface ore body, but the impression was gained that the breccia had largely disappeared before the more steeply dipping fault was reached. A gently dipping gouge seam curves into the main fault at the point of change in slope of the shaft, 50 feet on the slope above the level

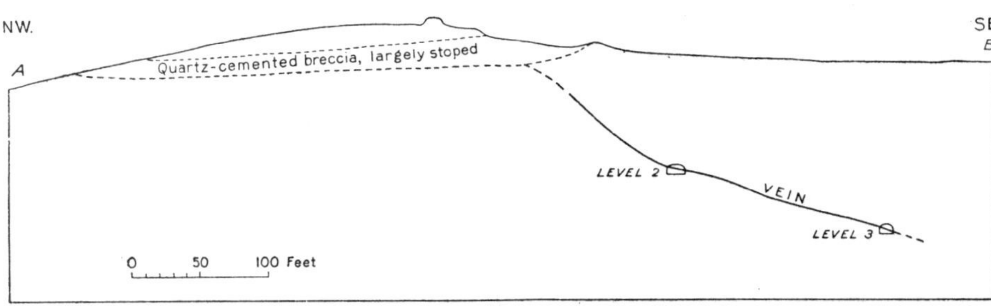

FIGURE 29.—Section through Blossom mine along line A–B, plate 52.

shown in plate 52. It would be expected that normal faulting would have occurred here, as in the Duplex and Good Hope mines, so that if there were any continuation of the ore it would be down the dip of this fault. Apparently no such continuation was found. A body of similar quartz-cemented breccia occurs on the hill east of the "fault" shaft (pl. 51) and might represent a continuation of the surface ore body if the fault were a reverse fault.

It is suggested that the peculiar flat vein represents the bottom of a roll in a vein originally more extensive but now in large part eroded. This vein may have had its "roots" or downward continuation toward the fault to the east, but the original relations are now obscured by postmineral movement. The peculiar relations of the surface ore body and the vein in the main shaft are obscure. However, both fractures were probably formed at about the same time with very little relative movement. The actual channel for the solutions to the breccia above was probably the vein revealed in the workings from the main shaft. This vein also contains flats and rolls. The flats appear to have been favorable to ore deposition and have been largely stoped.

The quartz-cemented breccia of the surface ore body is 10 feet or more thick, although, according to Ransome,[43] the actual ore had a

[43] Ransome, F. L., op. cit., p. 74.

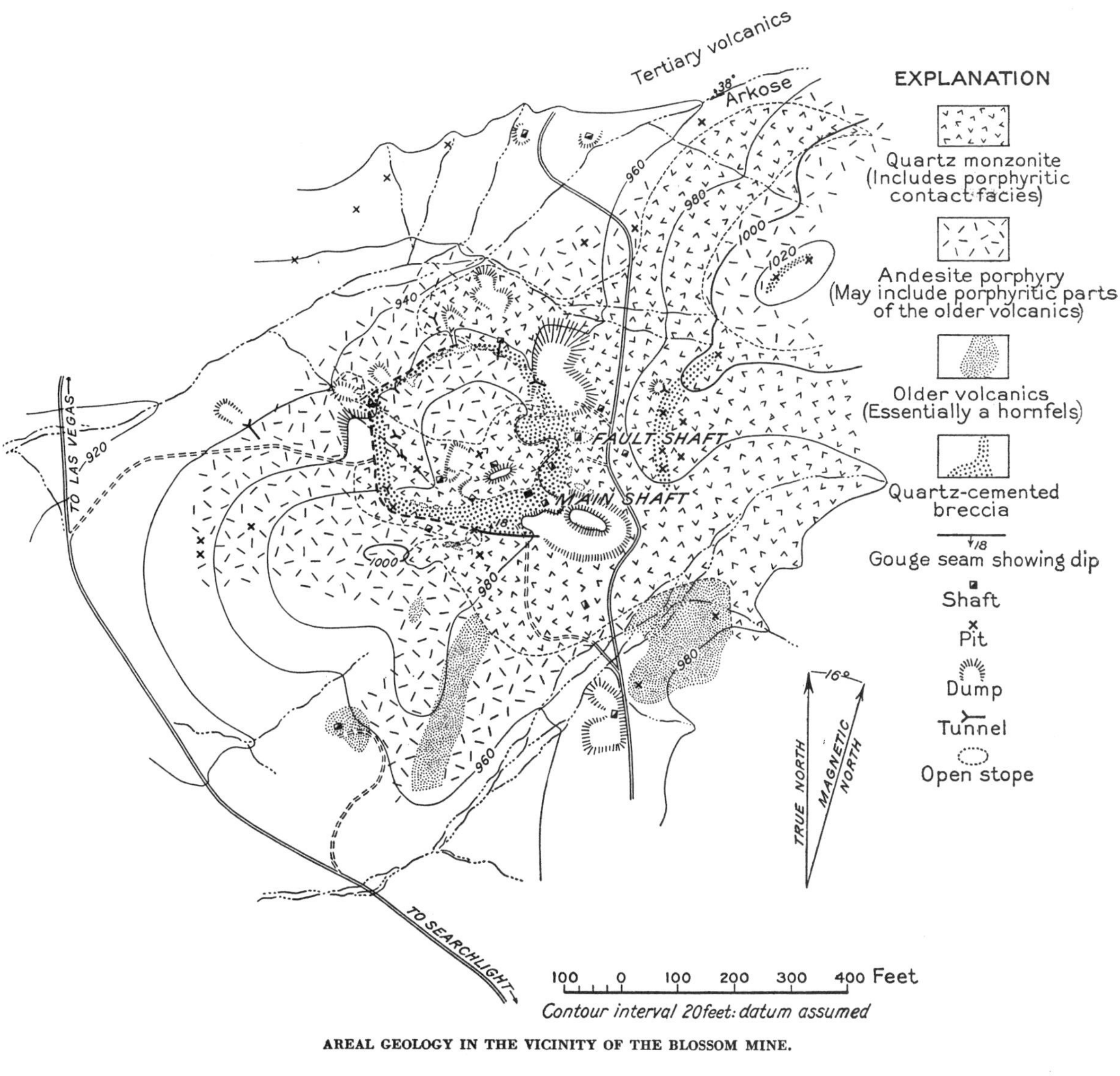

AREAL GEOLOGY IN THE VICINITY OF THE BLOSSOM MINE.

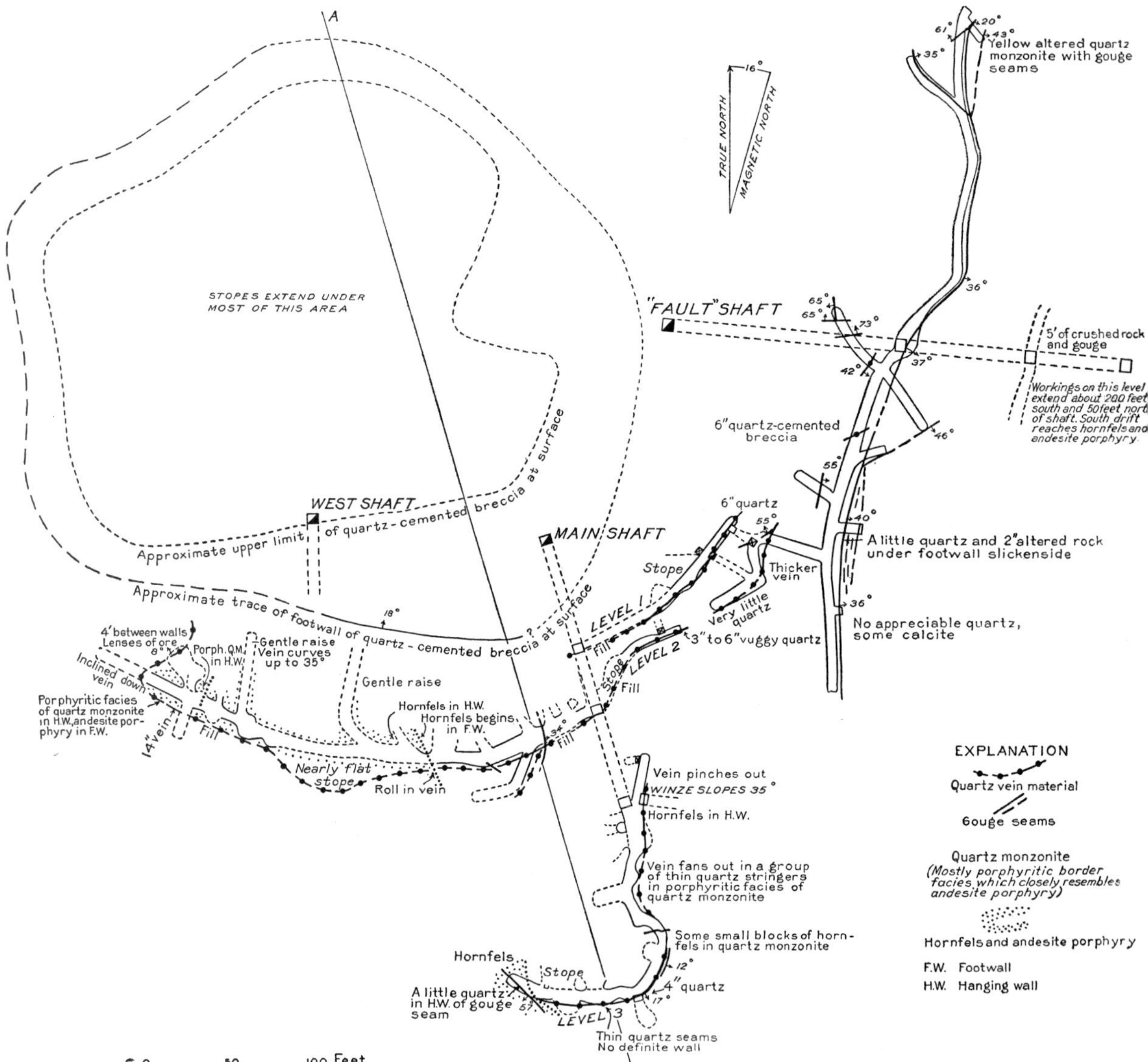

PLAN OF PRINCIPAL UNDERGROUND WORKINGS OF THE BLOSSOM MINE AND OUTLINE OF SURFACE ORE BODY.

From compass and tape traverse.

maximum thickness of 3 feet. The ore consisted of fragments of silicified country rock and porous sugary quartz. Gold was visible in some of this material, and the proportion of gold to silver, though variable, was commonly 1 to 4. Free gold was observed associated with limonite and small streaks of specular hematite. Ore that was being removed for shipping from the west end of level 2 in 1934 consisted of a breccia of altered country rock and radiating aggregates of vuggy comb quartz. White and gray clay and a little late calcite occur in the vugs and openings. Lumps of massive calcite with a fibrous structure are included in the ore and probably represent primary vein calcite. A few fragments show a little copper stain. The quartz tends to be shattered, apparently by late postmineral movements, so that much of it can be mined with a pick without recourse to blasting. Many of the stopes from the main shaft are 4 or 5 feet wide, but vein matter remaining in pillars is mostly between 1 and 2 feet wide. The ore body at the west end of level 2 is 4 feet wide between walls, but ore is restricted to lenses within this zone. On level 3 the vein fans out in a group of thin quartz seams without a definite wall.

A shaft on the top of the hill northwest of the main shaft is inclined from 62°–66° N. 10° W. for over 120 feet. It follows a fracture containing about 6 inches of gouge and a little quartz. The workings on the surface ore body are 28 feet below the collar of the shaft.

Further prospecting both above and below level 2 in the western extension of the vein may yield encouraging results. The fanning out of the vein on level 3 is not encouraging to deeper prospecting, and the pinching out of the vein to the northeast does not encourage prospecting in that direction. The zones of quartz-cemented breccia east of the "fault" shaft have been prospected to some extent but may still contain a streak of workable ore.

POMPEII

The Pompeii mine is about 1,500 feet north of the Blossom mine, in the northeastern part of sec. 22, T. 28 S., R. 63 E. It is credited with production in 1909, 1913–18, 1927–30, and 1933–34. Some milling was being done in 1931, and some ore was treated in 1933. The total production has been small. According to Ransome,[44] the shaft had been sunk to a depth of 335 feet at an angle of 58°, and one level had been driven at 265 feet by 1906. Water was abundant below the 265-foot level, and 65,000 gallons a day was being pumped. The water was used for domestic purposes, as well as for milling. A considerable surface establishment includes a 10-stamp mill, hoist, pumping equipment, and buildings.

[44] Ransome, F. L., op. cit., p. 74.

According to measurements made in 1934, the shaft is inclined 58°–59° about S. 30° W. There are levels at 41, 66, 124, 156, and 260 (the same as the 265-foot level) feet below the collar, and water stood 19 feet below the lowest level. The strike of the vein is about N. 65° W., and the dip though variable is about the same as that of the shaft. Considerable stoping has been done near the surface, and small stopes have been excavated on the 265-foot level, but the levels do not appear to extend far from the shaft.

The vein material on the upper levels consists of a breccia of yellow altered porphyritic igneous rock, probably a finely porphyritic facies of the andesite porphyry, cemented by thin veinlets of fine-grained, slightly vuggy quartz. Some of the quartz is platy, indicating former lamellar calcite. Thin sections show that the rock associated with the quartz is largely altered to adularia, and some adularia occurs with the quartz. Stains of iron oxide and black oxide of manganese occur in some of the veinlets, and, according to Ransome,[45] these veinlets contained the gold and silver. The 265-foot level reveals intersecting veins and seams that contain lenses of calcite and quartz. The microscope shows two generations of carbonate minerals—an early lamellar calcite with very fine-grained quartz and a later carbonate. Stains of iron oxide also are present in this material. No assays were available.

A prominent vein 1,500 feet north of the Pompeii shaft has been prospected by several shafts, adits, and pits and has been locally stoped. It strikes N. 41° E. and dips 30°–70° SE. A pit to the northeast of these workings east of the wash reveals a vein with similar strike but dipping 87° NW. The vein material consists of a breccia of altered and partly silicified country rock cemented with quartz and carbonate. Andesite porphyry occurs on both footwall and hanging wall.

A shaft 1,500 feet northeast of the Pompeii shaft, near the wash, reveals a vein 4 feet wide striking N. 70° W. and dipping 57° SW.

J. E. T.

The J. E. T. property consists of 9 claims centering about a group of workings 3,500 feet east of the Pompeii shaft, in sec. 28, T. 28 S., R. 63 E. The mine is credited with production in 1929. Probably most of the work was done in the early days of the camp, but the claims were relocated in 1927 by the present owners. The property is equipped with a 10-stamp mill. The main shaft is said to yield 7,000 gallons of water a day.

Three veins striking nearly east-west are exposed on this group of claims, and their relations are shown by the map (pl. 41) and in more detail in the sketch (fig. 30). Most of the work has been done on the

[45] Ransome, F. L., op. cit., p. 74.

northernmost of the three veins, which is exposed in shafts and pits for about 600 feet. It strikes N. 80° E. and dips mostly 43°–45° S.

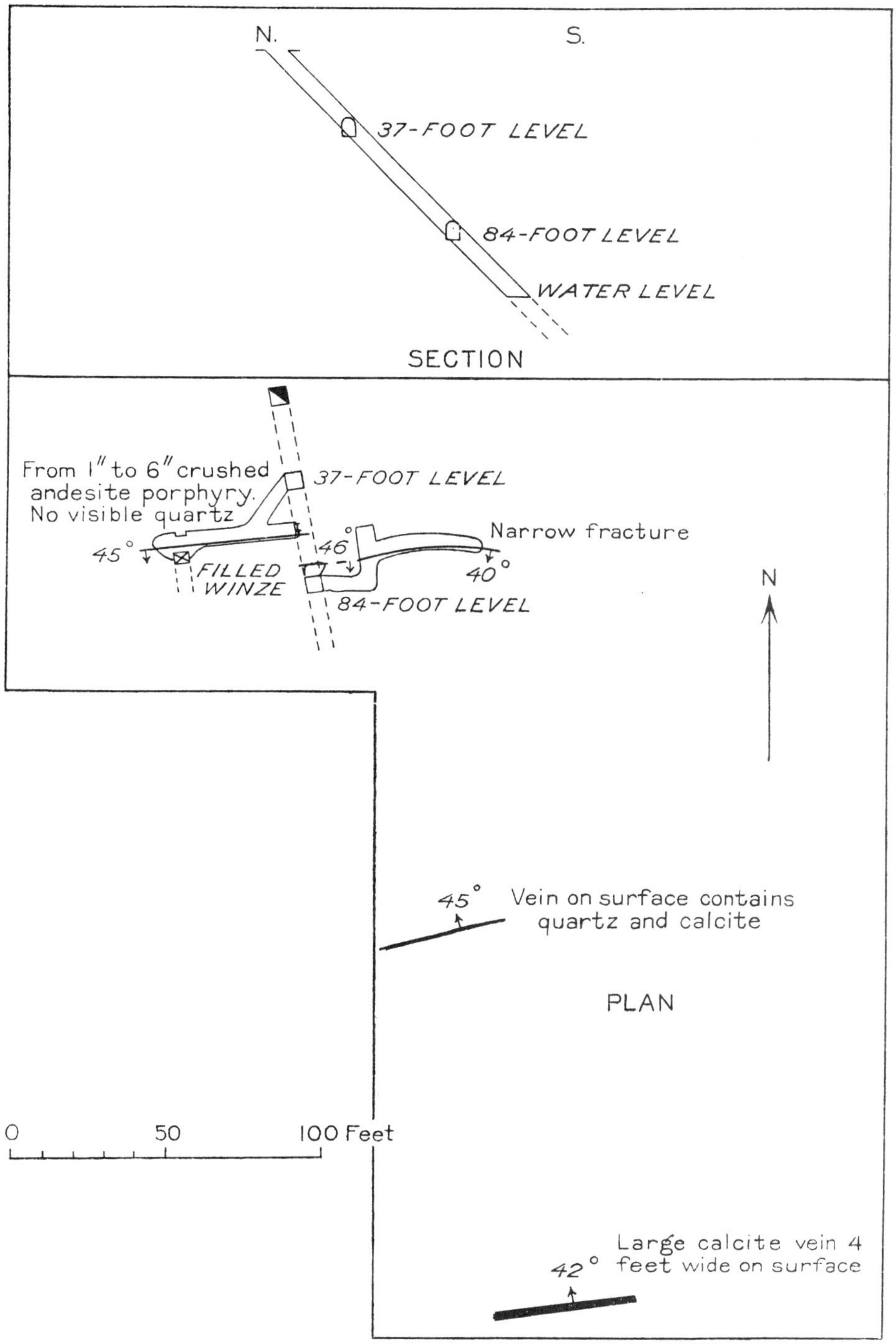

FIGURE 30.—Plan and section along shaft of J. E. T. workings. From compass and tape traverse.

The principal shaft shown in figure 30 extends 110 feet to water level on a 45° slope and to an unknown depth below. The country rock is andesite porphyry. Porphyry within the vein is brecciated. A very little quartz and some calcite appear in the vein. Some small spots are reported to have yielded high assays, but most of the assays recorded are low (0.1–0.4 ounce of gold to the ton). According to an engineer's report, assays revealed no gold below a depth of 90 feet in the shaft.

Two carbonate veins dipping 40°–45° N. and striking nearly east lie 180 and 300 feet south of the main vein. The southern vein is particularly prominent and is exposed for 1,000 feet. The carbonate vein in many places is 4 feet wide and contains coarse calcite, greenish

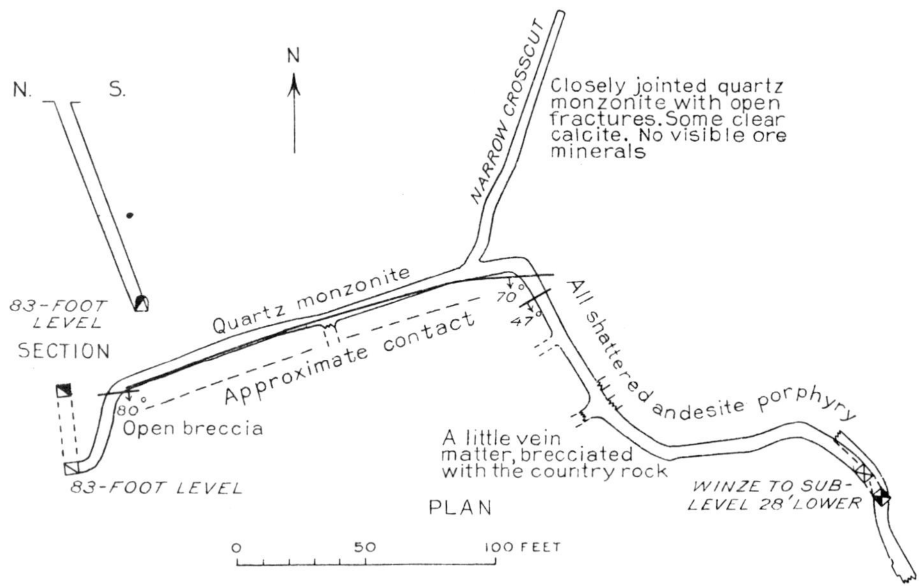

FIGURE 31.—Plan and section along shaft, Berlock workings. From compass and tape traverse.

or whitish clay, and very fine-grained quartz. Quartz in some of the material appears to have been brecciated and cemented with pinkish calcite. This material is reported to assay only a trace of gold.

It appears that the chief hope for production lies in the discovery of pockets or small bodies of gold ore sufficiently rich to mine, which may be found on the southward-dipping vein.

OTHER MINES AND PROSPECTS IN THE MAIN PART OF THE DISTRICT

A shaft known as the Berlock, 2,200 feet northeast of the Rambler workings, on the pediment slope east of the ridge, extends 83 feet down to a level (fig. 31) which is close to the contact of the quartz monzonite and a loose shattered breccia of andesite porphyry. A little quartz-vein material was found in the brecciated andesite porphyry. No production is known to have been obtained.

A shaft 1,400 feet east-northeast of the Good Hope shaft exposes a vertical vein 2 feet wide striking N. 15° E.

A vein 1,000 feet southeast of the Fraction shaft of the Duplex mine, as revealed by a shaft and pits, strikes N. 58° W. and dips 58° SW. at the shaft. No ore was found.

A group of shafts in quartz monzonite 1,400 feet east-northeast of the vein last described were sunk on stringers of quartz with a little copper stain trending nearly east-west and dipping 40°–70° S.

A shaft at the quartz monzonite contact 1,100 feet east-northeast of the Fraction shaft of the Duplex mine reveals a vein trending N. 80° E. and dipping 60° S. A little vuggy quartz occurs in both the quartz monzonite and the intruded hornfels.

Two nearly parallel breccia zones trending N. 78° W. and dipping 68° SW. are exposed by pits and an adit on the north side of the hill in the western part of Searchlight, near the highway.

A shaft 1,000 feet southeast of the Southern Nevada mine is inclined 75° S. 15° W. at the top. Material on the dump contains breccia with quartz veinlets and copper stain.

A shaft 1,000 feet west of the Searchlight M. & M. mine extends 173 feet to a crosscut and 40 feet farther to the bottom. The crosscut was driven 400 feet N. 14° E. in yellowish altered andesite porphyry. The Searchlight M. & M. vein, which it apparently was designed to intersect, was not found. The depth from surface to bedrock in this shaft is 120 feet.

Another shaft 700 feet southwest of that last described is 205 feet deep and is in alluvium throughout. A drift 33 feet to the east cuts bedrock of andesite porphyry. The bedrock surface slopes steeply to the west. Shafts farther west show yellowish altered porphyritic andesite, and those farthest to the west expose red porphyritic andesite, which probably belongs to the later group of volcanic rocks.

A shaft at the west side of the highway 1,000 feet north of the Searchlight M. & M. mine explores a vein containing as much as 8 inches of quartz which strikes N. 80° E. and dips 70° S. Another vein 400 feet southeast of this one strikes N. 80° E. and is explored by three vertical shafts and several pits.

MINES AND PROSPECTS IN THE EASTERN PART OF THE DISTRICT

The mines and prospects in the eastern part of the Searchlight district are not shown on plate 41 and were visited only briefly. The Boston mine in this area was not seen.

BOSTON

The Boston claims are about 1¼ miles northeast of Searchlight on the line between secs. 25 and 26, T. 18 S., R. 63 E., and were among the earliest to be prospected in the district. The following description is quoted from Ransome's report: [46]

> The Boston group of the Quartette Mining Co. was extensively prospected some years ago, but the attempt to find workable ore bodies was unsuccessful. There are two shafts on the property, an incline about 300 feet deep and a well-constructed vertical shaft 400 feet deep. Both are now (1906) dismantled. The workings are close to the eastern border of the quartz monzonite, in much shattered and greatly altered andesitic porphyry, which breaks readily into small dark-green or dark-red fragments and contains seams of epidote. The vein does not outcrop distinctly but apparently strikes nearly east and west. Some small bunches of good ore were found, consisting of chalcocite, malachite, azurite, chrysocolla, and cerusite, but an average value of about $6 per ton was the best that could be obtained for any considerable part of the deposit.

VIRGINIAN

The Virginian prospect is a short distance north of the Kingman Ferry road 2.1 miles east of Searchlight. A vein trending about N. 80° E. and dipping about 60° S. is prospected for a length of 400 feet by pits and shallow shafts. Some additional workings are offset from each end of the vein. The country rock is andesite porphyry. The vein material consists of quartz and fragments of country rock mostly less than 1 foot wide.

CHIEF OF THE HILLS

The Chief of the Hills mine is on a ridge known as Fourth of July Mountain, south the Kingman Ferry road 3 miles east of Searchlight. It was one of the properties of the Quartette Mining Co., and was operated for several years by lessees. Production is recorded for the years 1909–12, when ore was sent to the Cyrus Noble mill, 1917–19 and 1922–23, when it was operated by the Riverview Gold Mining & Milling Co. The total production has not been large. A mill and other buildings in apparent good condition stand on the west slope of the ridge.

The vein strikes about N. 85° W., or nearly at right angles to the ridge, so that the vein has been developed by adits rather than shafts. The lower drift is reached by a crosscut at the mill level, which was not accessible to the writer in 1931. An upper adit 93 feet long follows the vein, which has been stoped both above and below. At the next level above a caved stope slopes 76° N. A vein diverging slightly from the main vein on the south interescts it on the top of the ridge. The country rock is all andesite porphyry, which is epidotized in a broad zone paralleling the vein. The vein material as shown in one of the adits consists of brecciated andesite porphyry cemented with vuggy comb quartz. Cavities are partly filled with gray calcite, which in some places has an orange cast. Some manganese stain indicates that the calcite is manganiferous. No assays of this material were available.

[46] Ransome, F. L., op. cit., p. 72.

BIG CASINO

The Big Casino mine is south of the Kingman Ferry road about 3½ miles east of Searchlight. Production is recorded for 1913, 1915–16, 1918–21, and 1931–33. Copper and lead as well as gold and silver were produced. In 1931 the mine had been relocated as the Monte Carlo, and some work was being done on the surface north of the main shaft. Underground workings were wholly inaccessible. The main shaft is inclined 60° S. 40° W. in the upper part. The country rock is chiefly andesite porphyry. A cut south of the shaft exposed a vein containing quartz stringers trending due west and dipping 60° S. Another vein containing 1 foot of porous quartz strikes N. 65° W. and dips 40° SW. A vein near the cabin strikes N. 40° W. and dips 54° SW.

A pile of unweathered ore on the dump from the main shaft contains fragments of partly silicified country rock with scattered grains of galena, chalcopyrite, and sphalerite in quartz. According to a statement by the operators,[47] concentrates contained 4.2 ounces of gold and 12.1 ounces of silver to the ton, 17.9 percent of lead, 5.5 percent of copper, and 19.4 percent of zinc.

NEW ERA

The six claims constituting the New Era property are near the Kingman Ferry road 5.2 miles east of Searchlight. A small production is recorded for 1922, 1925, and 1931–34. The north shaft exposed a vein in andesite porphyry trending N. 45° W. and dipping 65° SW. The vein material is thoroughly weathered and consists of brecciated country rock and quartz without well-defined walls. The main shaft is inclined 65° N. 65° E. and is said to be 240 feet feep, with water level 200 feet below the surface.

---

[47] Weed, W. H., Mines Handbook, vol. 15, p. 1145, 1922.

INDEX

U.S. Geological Survey Mining Bulletins Reprint Series

These bulletins, issued early in this century, are much in demand as primary reference material for Nevada mining and water sources. All cover extensively mineralogy, topography, history and the various mines. All are 6 x 9 quality paperbacks with pages sewn together.

MINES OF THE GOLDFIELD, BULLFROG AND OTHER SOUTHERN NEVADA DISTRICTS, by F. L. Ransome, 144 pp, maps, index. Color map 22 x 28 laid in. This 1907 Bulletin also covers Searchlight, Eldorado, Crescent and Gold Mountain, The last 42 pages reproduce from Out West magazine two contemporary illustrated articles. "The Nevada Bonanzas of Today" and "Goldfield and the Goldfielders." These capture the flavor and excitement of Nevada's mining era.

MINES OF THE SILVER PEAK RANGE, KAWICH RANGE AND OTHER SOUTHWESTERN NEVADA DISTRICTS, by S.H. Ball. 218 pp, maps, index. Color map 22 x 28 laid in. This 1907 Bulletin also covers Lida-Goldfield Valley, Pahute Mesa, Cactus Range, Reveille Range, Belted Range, Tolicha Peak, the Amargosa Mountain system, Gold Mountain, Death Valley and the Panamint Range.

320 DESERT WATERING PLACES IN SOUTH-EASTERN CALIFORNIA AND SOUTHWESTERN NEVADA, by W. C. Mendenhall. 104 pp, illus, index. Color map 14 x 20 laid in. This 1909 Water-Supply Paper covers the Death Valley Basin in Inyo County and the Salton Sink of Imperial County, and numerous places in between. There are descriptions of 323 springs, all numbered and keyed to the colored map. The bulk of the book concerns San Bernardino County; in Nevada the following counties: Esmeralda, Nye, and Clark.

MINES OF BATTLE MOUNTAIN, REESE RIVER, AURORA AND OTHER WESTERN NEVADA DISTRICTS, by J. M. Hill. 200 pp, maps, index. This 1915 Bulletin also covers California's Lassen and Modoc counties, Douglas County, the Peavine district, and in Mineral County the Hawthorne, Granite, Santa Fe, Silver Star, and Pine Grove in Lyon County.

MINES OF TUSCARORA, CORTEZ AND OTHER NORTHERN NEVADA DISTRICTS, by W. H. Emmons. 220 pp, maps, index. This 1910 Bulletin also covers Midas, Good Hope, Columbia, Edgemont, Mountain City, Bullion, Safford, Tenabo, Lander, and the Lewis districts. Included is F. L. Ransome's "Notes on Some Mining Districts in Humboldt County" [Pershing County] 1909. Districts included are: Seven Troughs, Rosebud, Red Butte, Humboldt Range, Pahute Range, Sonoma Range.

MINES OF CHERRY CREEK, BRISTOL AND OTHER EASTERN NEVADA DISTRICTS, by J. M. Hill. 214 pp, maps, index. This 1916 Bulletin also includes Gold Butte, the Ruby Range, Mud Springs, Spruce Mountain, Mizpah, Kinsley, Toano Range, Tecoma, Ravenswood, Atlanta, Patterson, Troy, Bald Mountain, Granite, Hunter, Ward, Aurum, Duck Creek, Taylor and Kern districts.

An Essential Mining Reference...

MINING DISTRICTS AND MINERAL RESOURCES OF NEVADA, by Francis C. Lincoln. 295 pp, maps, index. This compilation gives a summary of each mining district in Nevada through 1923. There are historical summaries and bibliographies of each mining area, all arranged by counties. Section Two describes Nevada's mineral resources by types. An essential tool for the mining researcher.

NEVADA MINING REPRINTS

by William O. Vanderburg

All feature information on past operations, nature of mineral deposits and geology.

MINES OF CLARK COUNTY, 96 pp, illus, map. This report originally issued in 1937, covers many southern Nevada districts: Alunite, Arden, Crescent, Eldorado Canyon, Gold Butte, Goodsprings, St. Thomas, Searchlight.

MINES OF HUMBOLDT AND PERSHING COUNTIES, 128 pp, illus, map. This report, originally published in 1936 and 1938, covers many north-central Nevada districts: Awakening, Dutch Flat, Golconda, Gold Run, McDermitt, National, Sulphur, Warm Spings, Arabia, Farrell, Gold Banks, Kennedy, Mill City, Placerites, Rabbit Hole, Rochester, Rosebud, Scossa, Seven Troughs, Star, Trinity, Unionville, and others.

MINES OF LANDER AND EUREKA COUNTIES, 160 pp, illus, map. This report originally issued in 1938 and 1939, covers many eastern Nevada districts: Battle Mountain, Big Creek, Bullion, Hilltop, Kingston, Lewis, McCoy, Reese River (Austin), Buckhorn, Cortez, Diamond, Eureka, Mineral Hill, Palisade, Union, and others.

MINES OF CHURCHILL AND MINERAL COUNTIES, 144 pp, illus, map. This report, originally issued in 1937 and 1940, covers many west-central Nevada districts: Bernice, Fairview, Jessup, La Plata, Wonder, Aurora, Broken Hills, Candelaria, Garfield, Marietta, Rawhide, Silver Star and others.

CAMELS IN NEVADA, by Douglas McDonald. 32 pages, illus. In Nevada camel pack trains hauled salt, wood and even freight, also aiding early surveyors. But the beasts also brought problems. Modern camel races in Virginia City are recounted. Color cover.

THE BIG BONANZA, by Dan DeQuille (William Wright). 488 pages, illus., with intro. by Oscar Lewis. Indexed. Subtitled "An authentic account of the discovery, history, and working of the Comstock Lode," the Big Bonanza by DeQuille covers every phase of the epic rise of Virginia City, especially the special technology required to work the deep silver mines. Color cover.

LOST LEGENDS OF THE SILVER STATE, by Gerald Higgs. 142 pages, illus., Obvious, better known Nevada stories are dutifully ignored. Included are stories of Mark Twain being thrown out of Nevada, Civil War days, the $100.00 boulder, etc.

HISTORY OF NEVADA, 1913, by Sam P. Davis. 2 vols. 1344 pages, 60 illus. Originally issued in 1913, this landmark history is a compilation of special treatises on Nevada geography, Indians, territorial life, law and crime, mining history, politics, and biographies of prominent early 20th century Nevadans. The new index has 6,000 subject entries.

NEVADA TOWNS & TALES, by S. W. Paher, ed. 2 vols. 224 pages ea. 8-1/2x11. Chapters focus on Nevada's economic, social and geographical factors. Other major sections discuss state emblems, gambling, politics, mining, business and casino entertainment. There is much material on ghost towns, prospecting, legends, early day women, ranching, native animals, industries, banking and commerce, railroads, atomic testing, transportation, etc. Indexed. Color cover. Vol. 1, North, Vol. 2, South.

MY ADVENTURES WITH YOUR MONEY, by George Graham Rice. 334 pages, illus. Here are the memoirs of get-rich-quick financing of central Nevada and Death Valley mines, with interesting anecdotal material. Author capitalized the stocks of Goldfield, Greenwater and Rawhide mines, listed them on the national exchanges, and reaped the profits until convicted of mail fraud in 1911. More than 110 illustrations complement the text.

VIRGINIA CITY'S INTERNATIONAL HOTEL, by Richard C. Datin. 49 pages, illus. Subtitled Elegance on C Street. This book recreates the history, the people and the splendor of Virginia City. The International Hotel typified it all, kings, financiers, President U.S. Grant, and queens of the footlights.

SKETCHES OF VIRGINIA CITY, N.T., by J. Ross Browne. 48 pages, illus. In 1860 the author commented extensively on the miners and their madness over minerals, the Chinese, the Indians, the stagecoach drivers, proprietors, barroom brawlers, etc. Charming, humorous cartoons of these appear in the book.

NEVADA, AN ANNOTATED BIBLIOGRAPHY, by S.W. Paher. 585 pages, 7x10 illus. Here is a researcher's guide and description of more than 2,544 books relating to the history and development of Nevada. The useful 74-page index has 3,550 subjects; listed are 33 other bibliographies referring the researcher to countless Nevada materials.

JULIA BULETTE AND THE RED LIGHT LADIES OF NEVADA, by Douglas McDonald. 32 pages, illus., map. Here is the best written historical sketch to date of Virginia City's famed prostitute who was murdered in 1867. An overview of Nevada prostitution occupies the last part of the book, augmented by interesting photographs.

BODIE...BOOM TOWN. GOLD TOWN! The last of California's Old-time Mining Camps by Douglas McDonald. 48 pages, illustrated, color cover, 7x10 format. Though Bodie was discovered in 1859, no significant mining was begun until phenomenally rich strikes were made in 1877. The height of the boom occurred during 1879-80, though the mines were still active until about 1920. This heavily illustrated book shows the mines and miners, street scenes, important buildings, the mill, bullion, and the crowds which made up this great old mining camp.

MARK TWAIN IN VIRGINIA CITY, NEVADA, by Mark Twain. 192 pages. Twain portrays the life and the people of Virginia City, including the mining litigation, breaking in of a horse, etc. He mined for silver, labored in a silver mill and worked as a reporter for Virginia City's Territorial Enterprise. Drawings and cartoons. Paper.
